T0176683

Illuminating Statistical Analysis
Using Scenarios and Simulations

Illuminating Statistical Analysis Using Scenarios and Simulations

Jeffrey E Kottemann Ph.D.

Published by John Wiley & Sons, Inc., Hoboken, New Jersey
Published simultaneously in Canada

For general information on our other products and services or for technical support, please contact our Customer Care Department within the United States at (800) 762-2974, outside the United States at (317) 572-3993 or fax (317) 572-4002.

Wiley also publishes its books in a variety of electronic formats. Some content that appears in print may not be available in electronic formats. For more information about Wiley products, visit our web site at www.wiley.com.

Library of Congress Cataloging-in-Publication Data:

Names: Kottemann, Jeffrey E.
Title: Illuminating statistical analysis using scenarios and simulations/
    Jeffrey E Kottemann, Ph.D.
Description: Hoboken, New Jersey: John Wiley & Sons, Inc. [2017], | Includes
    index.
Identifiers: LCCN 2016042825| ISBN 9781119296331 (cloth) | ISBN 9781119296362
    (epub)
Subjects: LCSH: Mathematical statistics. | Distribution (Probability theory)
Classification: LCC QA276 .K676 2017 | DDC 519.5–dc23 LC record available at
    https://lccn.loc.gov/2016042825

Printed in the United States of America

10 9 8 7 6 5 4 3 2 1

Table of Contents

Preface

The goal of this book is to help people develop an assortment of key intuitions about statistics and inference and use those intuitions to make sense of statistical analysis methods in a conceptual as well as a practical way. Moreover, I hope to engender good ways of thinking about uncertainty. The book is comprised of a series of short, concise chapters that build upon each other and are best read in order. The chapters cover a wide range of concepts and methods of classical (frequentist) statistics and inference. (There are also appendices on Bayesian statistics and on data mining techniques.)

Examining computer simulation results is central to our investigation. Simulating pollsters, for example, who survey random people for responses to an agree or disagree opinion question not only mimics reality but also has the added advantage of being able to employ 1000 independent pollsters simultaneously. The results produced by such simulations provide an eye-opening way to (re)discover the properties of sample statistics, the role of chance, and to (re)invent corresponding principles of statistical inference. The simulation results also foreshadow the various mathematical formulas that underlie statistical analysis.

Mathematics used in the book involves basic algebra. Of particular relevance is interpreting the relationships found in formulas. Take, for example, $w = x/y$. As x increases, w increases because x is the numerator of the fraction. And as y increases, w decreases because y is the denominator. Going one step further, we could have $w = x/(y/z)$. Here, as z increases, (y/z) decreases, $x/(y/z)$ increases, so w increases. These functional forms mirror most of the statistical formulas we will encounter.

As we will see for a wide range of scenarios, simulation results clearly illustrate the terms and relationships found in the various formulas that underlie statistical analysis methods. They also bring to light the underlying assumptions that those formulas and methods rely upon. Last, but not least, we will see that simulation can serve as a robust statistical analysis method in its own right.

Bon voyage
Jeffrey E. Kottemann

Acknowledgements

My thanks go to Dan Dolk, Gene Hahn, Fati Salimian, and Kathie Wright for their feedback and encouragement. At John Wiley & Sons, thanks go to Susanne Steitz-Filler, Kathleen Pagliaro, Vishnu Narayanan, and Shikha Pahuja.

Part I
Sample Proportions and the Normal Distribution

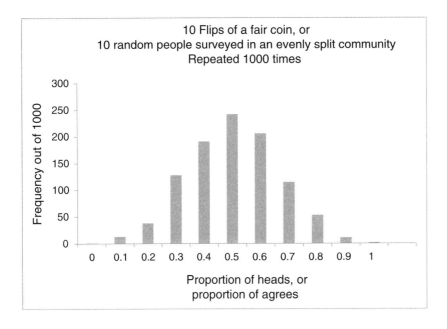

Illuminating Statistical Analysis Using Scenarios and Simulations, First Edition.
Jeffrey E Kottemann.
© 2017 John Wiley & Sons, Inc. Published 2017 by John Wiley & Sons, Inc.

1

Evidence and Verdicts

Before we focus in on using statistics as evidence to be used in making judgments, let's take a look at a widely used "verdict outcomes framework." This general framework is useful for framing judgments in a wide range of situations, including those encountered in statistical analysis.

Anytime we use evidence to arrive at a judgment, there are four generic outcomes possible, as shown in Table 1.1. Two outcomes correspond to correct judgments and two correspond to incorrect judgments, although we rarely know whether our judgments are correct or incorrect. Consider a jury trial in U.S. criminal court. Ideally, the jury is always correct, judging innocent defendants not guilty and judging guilty defendants guilty. Evidence is never perfect, though, and so juries will make erroneous judgments, judging innocent defendants guilty or guilty defendants not guilty.

Table 1.1 Verdict outcomes framework.

The verdict / Unknown Truth	Not guilty verdict (do not reject presumption)	Guilty verdict (reject presumption)
Defendant is innocent (the presumption)	Correct	Incorrect verdict (type I error)
Defendant is guilty	Incorrect verdict (type II error)	Correct

In U.S. criminal court, the presumption is that a defendant is innocent until "proven" guilty. Further, convention in U.S. criminal court has it that we are more afraid of punishing an innocent person (type I error) than we are of letting a guilty person go unpunished (type II error). Because of this fear, the threshold for a guilty verdict is set high: "Beyond a reasonable doubt." So, convicting an

Illuminating Statistical Analysis Using Scenarios and Simulations, First Edition.
Jeffrey E Kottemann.

innocent person should be a relatively unlikely outcome. In U.S. criminal court, we are willing to have a greater chance of letting a guilty person go unpunished than we are of punishing an innocent person. In short, we need to be very sure before we reject the presumption of innocence and render a verdict of guilty in U.S. criminal court.

We can change the relative chances of the two types of error by changing the threshold. Say we change from "beyond a reasonable doubt" to "a preponderance of evidence." The former is the threshold used in U.S. criminal court, and the latter is the threshold used in U.S. civil court. Let's say that the former corresponds to being 95% sure before judging a defendant guilty and that the latter corresponds to being 51% sure before judging a defendant guilty. You can imagine cases where the same evidence results in different verdicts in criminal and civil court, which indeed does happen. For example, say that the evidence leads to the jury being 60% sure of the defendant's guilt. The jury verdict in criminal court would be not guilty (60% < 95%) but the jury verdict in civil court would be guilty (60% > 51%). Compared to criminal court, civil court is more likely to declare an innocent person guilty (type I error), but is also less likely to declare a guilty person not guilty (type II error).

Changing the verdict threshold
either decreases type I error while increasing type II error (criminal court)
or increases type I error while decreasing type II error (civil court)

Statistical analysis is conducted as if in criminal court. Below are a number of jury guidelines that have parallels in statistical analysis, as we'll see repeatedly.

Erroneously rejecting the presumption of innocence (type I error) is feared most.

The possible verdicts are "guilty" and "not guilty." There is no verdict of "innocent."

Reasons other than perceived innocence can lead to a not guilty verdict, such as insufficient evidence.

To reject the presumption of innocence and render a guilty verdict, there must be a sufficient amount of (unbiased) evidence.

To reject the presumption of innocence and render a guilty verdict, the pieces of evidence must be sufficiently consistent (not at variance with each other).

Statistical analysis formally evaluates evidence in order to determine whether to reject or not reject a stated presumption, and it is primarily concerned with limiting the chances of type I error. Further, the amount of evidence and the variance of evidence are key characteristics of evidence that are formally incorporated into the evaluation process. In what follows, we'll see how this is accomplished.

2

Judging Coins I

Let's start with the simplest statistical situation: that of judging whether a coin is fair or not fair. Later we'll see that this situation is statistically equivalent to agree or disagree opinion polling. A coin is fair if it has a 50% chance of coming up heads, and a 50% chance of coming up tails when you flip it. Adjusting the verdict table to the coin-flipping context gives us Table 2.1.

Table 2.1 Coin flipping outcomes.

Your verdict / Unknown truth	Coin is fair (do not reject presumption)	Coin is not fair (reject presumption)
Coin is fair (the presumption)	Correct	Incorrect: type I error
Coin is not fair	Incorrect: type II error	Correct

Statistical Scenario–Coins #1

You want to see if a coin is fair or not. To gather evidence, you plan on flipping it 10 times to see how many heads come up. Beforehand, you want to set the verdict rule.

Where do you draw the two lines for declaring the coin to be not fair?

0   1   2   3   4   5   6   7   8   9   10
Number of heads

Illuminating Statistical Analysis Using Scenarios and Simulations, First Edition.
Jeffrey E Kottemann.
© 2017 John Wiley & Sons, Inc. Published 2017 by John Wiley & Sons, Inc.

Draw one line toward the left for your threshold of "too few heads," and another line toward the right for your threshold of "too many heads."

Where you draw the lines represents *your* choice of thresholds.

Just use your intuition; don't do any arithmetic.

Intuitively, it seems extremely unlikely for a fair coin to come up heads only 0 or 1 times out of 10, and most people would arrive at the verdict that the coin is not fair. Likewise, it seems extremely unlikely for a fair coin to come up heads 9 or 10 times out of 10, and most people would arrive at the verdict that the coin is not fair. On the other hand, it seems fairly likely for a fair coin to come up heads 4, 5, or 6 times out of 10, and so most people would say that the coin seems fair. But what about 2, 3, 7, or 8 heads? Let's experiment.

Shown in Figure 2.1 is a histogram of what actually happened (in simulation) when 1000 people each flipped a *fair coin* 10 times. This shows us how fair coins tend to behave. The horizontal axis is the number of heads that came up out of 10. The vertical axis shows the number of people out of the 1000 who came up with the various numbers of heads.

Appendix B gives step-by-step instructions for constructing this simulation using common spreadsheet software; guidelines are also given for constructing additional simulations found in the book.

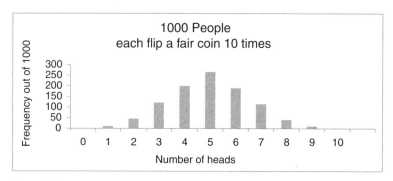

Figure 2.1

Sure enough, fair coins very rarely came up heads 0, 1, 9, or 10 times. And, sure enough, they very often came up heads 4, 5, or 6 times. What about 2, 3, 7, or 8 heads?

Notice that 2 heads came up a little less than 50 times out of 1000, or near 5% of the time. Same with 8 heads. And, 3 heads came up well over 100 times out of 1000, or over 10% of the time. Same with 7 heads.

Where do you want to draw the threshold lines now?

0   1   2   3   4   5   6   7   8   9   10
                  Number of heads

How about the lines shown on the histogram in Figure 2.2?

Eyeballing the histogram, we can add up the frequencies for 0, 1, 2, 8, 9, and 10 heads out of the 1000 total. Fair coins came up with 0, 1, 2, 8, 9, or 10 heads a total of about $0 + 10 + 45 + 45 + 10 + 0 = 110$ and $110/1000 = 11\%$ of the time.

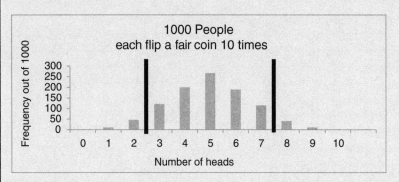

Figure 2.2

So, a "verdict rule" that has about an 11% chance of type I error can be stated as:

If the number of heads flipped is *outside* the interval

Number of heads ≥ 3 and number of heads ≤ 7

Then reject the presumption that the coin is fair.

Fair coins will be inside the interval about 89% of the time, and we will correctly judge them to be fair coins.

Fair coins will be outside the interval about 11% of the time, and we will incorrectly judge them to be unfair coins.

The "verdict rule" defines an interval outside of which we will reject the presumption.

Type I error   occurs when a fair coin's head count is outside our interval.
Type II error occurs when an unfair coin's head count is inside our interval.

3

Brief on Bell Shapes

Statistical Scenario—Distribution of Outcomes.

Why is the histogram bell-shaped?

Before expanding the previous *Statistical Scenario* let's briefly explore why the histogram, reproduced in Figure 3.1, is shaped the way it is: bell-shaped. It tapers off symmetrically on each side from a single peak in the middle.

Since each coin flip has two possible outcomes and we are considering ten separate outcomes together, there are a total of $2^{10} = 1024$ unique possible patterns (permutations) of heads and tails with 10 flips of a coin. Of these, there is only one with 0 heads and only one with 10 heads. These are the least likely outcomes.

TTTTTTTTTT   HHHHHHHHHH

There are ten with 1 head, and ten with 9 heads:

HTTTTTTTTT	THHHHHHHHH
THTTTTTTTT	HTHHHHHHHH
TTHTTTTTTT	HHTHHHHHHH
TTTHTTTTTT	HHHTHHHHHH
TTTTHTTTTT	HHHHTHHHHH
TTTTTHTTTT	HHHHHTHHHH
TTTTTTHTTT	HHHHHHTHHH
TTTTTTTHTT	HHHHHHHTHH
TTTTTTTTHT	HHHHHHHHTH
TTTTTTTTTH	HHHHHHHHHT

Illuminating Statistical Analysis Using Scenarios and Simulations, First Edition.
Jeffrey E Kottemann.
© 2017 John Wiley & Sons, Inc. Published 2017 by John Wiley & Sons, Inc.

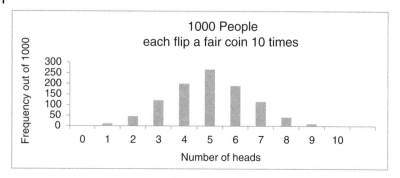

Figure 3.1

Since there are 10 times more ways to get 1 or 9 heads than 0 or 10 heads, we expect to flip 1 or 9 heads 10 times more often than 0 or 10 heads.

Further, there are 45 ways to get 2 or 8 heads, 120 ways to get 3 or 7 heads, and 210 ways to get 4 or 6 heads. Finally, there are 252 ways to get 5 heads, which is the most likely outcome and therefore the most frequently expected outcome. Notice how the shape of the histogram of simulation outcomes we saw in Figure 3.1 closely mirrors the number of ways (#Ways) chart that is shown in Figure 3.2.

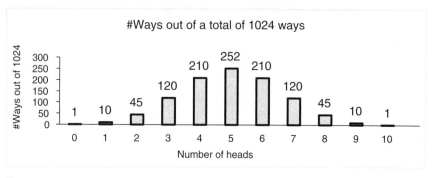

Figure 3.2

You don't need to worry about calculating #ways. Soon we won't need such calculations. Just for the record, the formula for the #ways is $n!/[h! \times (n - h)!]$ where n is the number of flips, h is the number of heads you are interested in, and ! is the factorial operation (example: $5! = 5 \times 4 \times 3 \times 2 \times 1 = 120$). In official terms, #ways is the number of combinations of n things taken h at a time.

4

Judging Coins II

Let's revisit *Statistical Scenario–Coins #1*, now with additional information on each of the possible outcomes. Table 4.1 summarizes this additional information. As noted, there are a total of $2^{10} = 1024$ different unique patterns of heads & tails possible when we flip a coin 10 times. For any given number of heads, as we have just seen, there are one or more ways to get that number of heads.

Statistical Scenario—Coins #2

Using Table 4.1, where do you draw the two lines for declaring the coin to be unfair?

Table 4.1 Coin flipping details.

#Heads	#Ways	Expected relative frequency	Probability	as Percent	Rounded
0	1	1/1024	0.00098	0.098%	0.1%
1	10	10/1024	0.00977	0.977%	1.0%
2	45	45/1024	0.04395	4.395%	4.4%
3	120	120/1024	0.11719	11.719%	11.7%
4	210	210/1024	0.20508	20.508%	20.5%
5	252	252/1024	0.24609	24.609%	24.6%
6	210	210/1024	0.20508	20.508%	20.5%
7	120	120/1024	0.11719	11.719%	11.7%
8	45	45/1024	0.04395	4.395%	4.4%
9	10	10/1024	0.00977	0.977%	1.0%
10	1	1/1024	0.00098	0.098%	0.1%
Totals:	1024	1024/1024	1.0	100%	100%

Illuminating Statistical Analysis Using Scenarios and Simulations, First Edition.
Jeffrey E Kottemann.
© 2017 John Wiley & Sons, Inc. Published 2017 by John Wiley & Sons, Inc.

The #ways divided by 1024 gives us the <u>expected</u> <u>relative frequency</u> for that number of heads expressed as a fraction. For example, we expect to get 5 heads $252/1024^{ths}$ of the time. The fraction can also be expressed as a decimal value. This decimal value can be viewed as the <u>probability</u> that a certain number of heads will come up in 10 flips. For example, the probability of getting 5 heads is approximately 0.246. We can also express this as a percentage, 24.6%.

A probability of 1 (100%) means something will always happen and a probability of 0 (0%) means something will never happen. A probability of 0.5 (50%) means something will happen half the time. The sum of the probabilities of the *entire set* of possible outcomes is the sum of all the probabilities and always equals 1 (100%). The probability of a *subset* of possible outcomes can be calculated by summing the probabilities of each of the outcomes. For example, using the rounded percentages from the table, the probability of *2 or fewer heads* is $0.1\% + 1.0\% + 4.4\% = 5.5\%$.

Notice how the bars of our simulation histogram, reproduced in Figure 4.1, reflect the corresponding probabilities in Table 4.1.

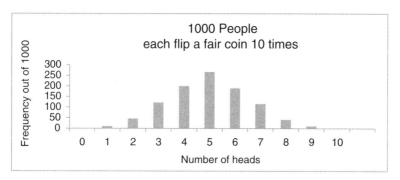

Figure 4.1

Say someone gives you a coin to test. When you flip the coin 10 times, you are sampling the coin's behavior 10 times. The number of heads you toss is your evidence. Based on this evidence you must decide whether to reject your presumption of fairness and judge the coin as not fair.

What happens if you make your "verdict rule" to be:

Verdict "coin is not fair" if outside the interval *#heads ≥ 1 and ≤ 9* as shown in Table 4.2 and the accompanying Figure 4.2?

Table 4.2 First verdict rule scenario.

Your verdict / Unknown truth	Coin is fair do not reject presumption *If #heads ≥ 1 and ≤ 9*	Coin is not fair reject presumption *If outside the interval*
Coin is fair the presumption	Correct	Incorrect: type I error *0.2% chance*
Coin is not fair	Incorrect: type II error	Correct

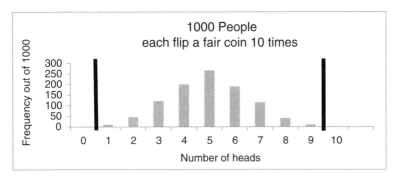

Figure 4.2

From the *Statistical Scenario* Table 4.1, we can see that a fair coin will come up 0 heads about 0.1% of the time, and 10 heads about 0.1% of the time. The sum is about 0.2% of the time, or about 2 out of 1000. So, it will be extremely rare for us to make a type I error and erroneously call a fair coin unfair because fair coins will almost never come up with 0 or 10 heads. However, what about 1 head or 9 heads? Our rule says not to call those coins unfair. But a fair coin will only come up 1 head or 9 heads about $1\% + 1\% = 2\%$ of the time. Therefore, we may end up misjudging many unfair coins that come up heads one or nine times because we'll declare them to be fair coins. That is type II error.

Determining the chance of type II error is too involved for discussion now (that is Chapter 17), but recall from Chapter 1 that increasing the chance of type I error decreases the chance of type II error, and vice versa.

To lower the chances of type II error, we can narrow our "verdict rule" interval to *#heads ≥ 2 and ≤ 8* as shown in Table 4.3 and Figure 4.3. Now the probability of making a type I error is about $0.1\% + 1\% + 1\% + 0.1\% = 2.2\%$. This rule will decrease the chances of type II error, while increasing the chances of type I error from 0.2 to 2.2%.

Table 4.3 Second verdict rule scenario.

Your verdict ⟍ Unknown Truth	Coin is fair do not reject presumption *If #heads ≥ 2 and ≤ 8*	Coin is not fair reject presumption *If outside the interval*
Coin is fair the presumption	Correct	Incorrect: type I error *2.2% chance*
Coin is not fair	Incorrect: type II error	Correct

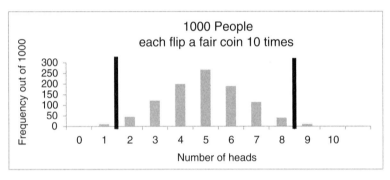

Figure 4.3

If we narrow our "verdict rule" interval even more to #heads ≥ 3 and ≤ 7, we get Table 4.4 and Figure 4.4.

Table 4.4 Third verdict rule scenario.

Your verdict ⟍ Unknown truth	Coin is fair do not reject presumption *If #heads ≥ 3 and ≤ 7*	Coin is not fair reject presumption *If outside the interval*
Coin is fair the presumption	Correct	Incorrect: type I error *11% chance*
Coin is not fair	Incorrect: type II error	Correct

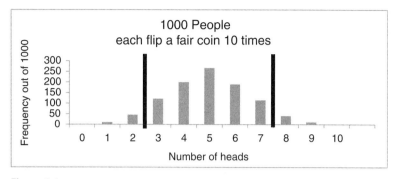

Figure 4.4

Now the probability of making a type I error is about $0.1\% + 1\% + 4.4\% + 4.4\% + 1\% + 0.1\% = 11\%$ because a fair coin will come up 0, 1, 2, 8, 9, or 10 heads about 11% of the time. We can express this uncertainty by saying either that there will be an 11% chance of a type I error, or that we are 89% confident that there will not be a type I error. Notice that this is what we came up with earlier by simply eyeballing the histogram of actual simulation outcomes in Chapter 2.

As noted earlier, type I error is feared most. And an 11% chance of type I error is usually seen as excessive. So, we can adopt this rule:

Verdict: If outside the interval #Heads ≥ 2 and ≤ 8, Judge Coin to be Not Fair.

This gives us about a 2% chance of type I error.

From now on, we'll typically use the following threshold levels for type I error: 10% (0.10), 5% (0.05), and 1% (0.01). We'll see the effects of using various thresholds as we go along. Also as we go along we'll need to replace some common words with statistical terminology. Below are statistical terms to replace the common words we have been using.

> *Alpha-Level* is the threshold probability we stipulate for the occurrence of type I error. Commonly used alpha-levels are 0.05 and 0.01. Alpha-levels used in sciences such as physics are typically much lower.
>
> *Confidence level* is the complement of alpha-level and is the threshold percentage we stipulate for the nonoccurrence of type I error. 95% confidence corresponds to an alpha-level of 0.05. 99% confidence corresponds to an alpha-level of 0.01.
>
> *Null hypothesis* is the presumption we have been referring to above. Example: the coin is not unusual, it is a fair coin. Most of the simulations in this book simulate outcomes to expect when the null hypothesis is true: like a fair coin.

Binomial is a variable with only two possible values, such as heads or tails. The term comes from the Latin for "two names." Other examples of binomials are agree or disagree, male or female. When we represent the binomial values as 0 or 1, then the average will give us the proportion of 1s. For example, the average of 0, 1, 0, and 1 is 0.5 and the average of 1, 1, 0, and 1 is 0.75. (Appendix A overviews all the types of data we'll be working with as the book progresses.)

Sample is the set of observations we have, which is the set of heads and tails flipped for the above coin flipping cases. A sample constitutes evidence.

Sample size is the number of separate observations, which equals 10 for the above cases. The italicized letter n is often used: the sample $n = 10$.

Sample statistic is a calculated value that serves to summarize the sample. They are summarizations of evidence. Examples: number of heads in the sample or Proportion of heads in the sample (number of heads divided by sample size). Counts and proportions are the basic sample statistics for binomial variables.

Sampling distributions are illustrated by the histograms of simulation results. They reflect the distributions of the values for sample statistics we get with repeated sampling.

It is important to emphasize that simulation histograms represent sampling distributions that tell us what to expect when the null hypothesis is true. We'll look at many, many sampling distribution histograms in this book. For the remainder of Part I, we'll switch from using counts as our sample statistic to using proportions as our sample statistic. The sampling distribution histogram in Figure 4.5 shows the use of proportions rather than counts on the horizontal axis.

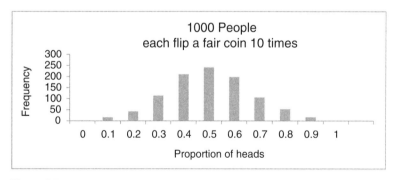

Figure 4.5

Null and Alternate Hypotheses

The null hypothesis is a statement that may be ruled out by evidence in favor of an alternate hypothesis. Typically, the null hypothesis is an equality and the alternate hypothesis is an opposing inequality. For example:

Null   The coin's true proportion of heads *is equal to* 0.5.

Alternate   The coin's true proportion of heads *is not equal to* 0.5.

From Evidence to Verdict

We choose our sample size ($n = 10$) and establish our verdict threshold ($alpha = 0.05$) and corresponding interval. We collect a sample of individual observations (flipped a head or a tail) and calculate the sample statistic of interest (number or proportion of heads). We then determine whether the sample statistic falls outside our interval. If so, we reject the null hypothesis (the presumption), otherwise we do not reject the null hypothesis. Officially, we never accept the null hypothesis, we either reject it or do not reject it. This is analogous to what we saw in court earlier: There is no verdict "innocent," the possible verdicts are either "guilty" or "not guilty."

Type I error: Rejecting the null hypothesis when it is actually true. Example: A fair coin happens to come up heads 10 out of 10 times.

Type II error: Not rejecting the null hypothesis when it is actually false. Example: An unfair coin happens to come up heads 5 out of 10 times.

5

Amount of Evidence I

<table>
<tr><td>

Statistical Scenario—Amount of Evidence I

What happens to the "verdict rule" interval when the amount of evidence increases?
</td></tr>
</table>

Let's start by considering very small amounts of evidence first: 1000 people flip a fair coin only *two times*. The possible outcomes are TT, TH, HT, and HH. So, we expect about 250 people to get 0% heads (TT) and about 250 people to get 100% heads (HH), which are both extremely far from the 50% heads (TH, HT) we expect with a fair coin. Next, recall that with *10 flips* such extremes are quite infrequent. With *100 flips*, then, such extremes should be almost unheard of. Let's look at three corresponding simulation histograms.

When there are only two flips ($n = 2$), half of the sample proportions are at the extremes of 0 or 1, as shown in Figure 5.1.

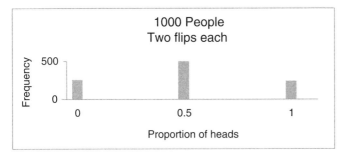

Figure 5.1

Compared to two flips, when 1000 people flip a fair coin 10 times ($n = 10$), there are fewer extreme sample proportions, as shown in Figure 5.2.

Illuminating Statistical Analysis Using Scenarios and Simulations, First Edition.
Jeffrey E Kottemann.
© 2017 John Wiley & Sons, Inc. Published 2017 by John Wiley & Sons, Inc.

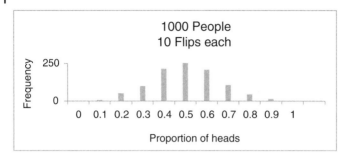

Figure 5.2

And when 1000 people flip a fair coin 100 times ($n = 100$), there are fewer still, as shown in Figure 5.3. The larger the sample size, the less likely it is that people will get sample proportions that are far from what we expect for a fair coin: 0.5.

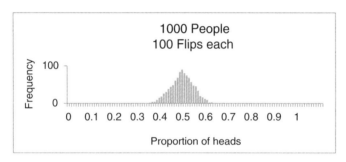

Figure 5.3

It stands to reason that when the sample proportion histograms become less spread out, then the "verdict rule" intervals should become less spread out too. Let's look at the difference in the "verdict rule" intervals for sample sizes of 10 versus 100. The "verdict rule" interval lines are placed so that both histograms have approximately the same chance of type I error.

Figure 5.4 is the histogram for sample size of 10.

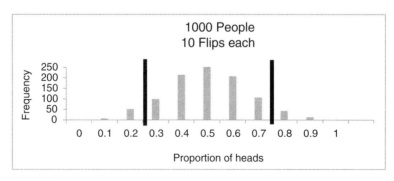

Figure 5.4

The "verdict rule" interval of 0.3–0.7 gives us close to a 10% chance of type I error *when sample size is 10*. About 900 of the results are contained in the interval 0.3–0.7, and about 100 are outside, 50 on each side.

Figure 5.5 is the histogram for sample size of 100, "zooming in" on the region around 0.5.

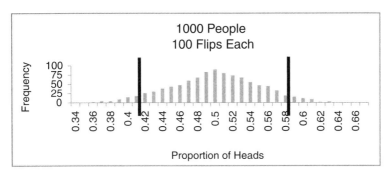

Figure 5.5

This smaller "verdict rule" interval of 0.42–0.58 also gives us close to a 10% chance of type I error *when sample size is 100*. About 900 of the results are contained in the interval 0.42–0.58, and about 100 are outside, 50 on each side.

As you can see:

> When our sample size is *larger* our "verdict rule" can use *smaller* intervals.

It is important to keep in mind that sample size and uncertainty are intimately related. As the sample size increases, the sample proportions tend to condense closer to the true proportion (0.5 with a fair coin). The uncertainty of the sample proportions decreases because we expect them to be closer to the true proportion as sample size increases.[1]

> More evidence (larger sample sizes) → less uncertainty

As we'll see later, the uncertainty due to sample size n is incorporated into various statistical formulas.[2]

---

1 This principle is called the Law of Large Numbers.
2 The sample proportion is often symbolized by $\hat{p}$ ("*p* hat"), but I will continue to spell it out.

6

Variance of Evidence I

Statistical Scenario—Variance of Evidence I

What happens to our "verdict rule" interval when the chance of heads is not 50%?

Let's next look at what happens when the coin is unbalanced and the chance of heads is more or less than 50%. As we'll see, a 50% coin will have more outcome variety than a 25% coin that, in turn, will have more outcome variety than a 10% coin. A coin with 0 or 100% chance of heads will have no outcome variety whatsoever. Variety decreases more and more as the chance of heads moves from 50% to 0% or 100% because variety becomes less and less likely. Let's look at three corresponding simulation histograms.

Figure 6.1 shows sample proportions with coins whose chance of heads is 50%.

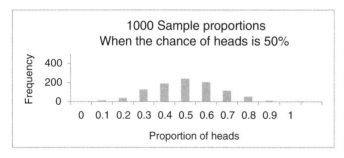

Figure 6.1

Figure 6.2 is for coins whose chance of heads is 25%; its outcome variety is less; it is more bunched up.

Illuminating Statistical Analysis Using Scenarios and Simulations, First Edition.
Jeffrey E Kottemann.
© 2017 John Wiley & Sons, Inc. Published 2017 by John Wiley & Sons, Inc.

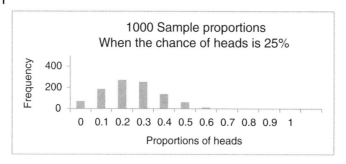

Figure 6.2

And Figure 6.3 is for coins whose chance of heads is 10%; its outcome variety is even less; it is even more bunched up (it is also lopsided, which we'll consider in Chapter 9).

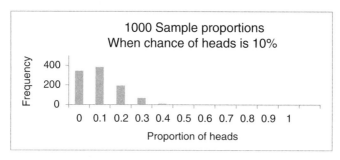

Figure 6.3

For coins whose chance of heads is 0%, all the sample proportions will be equal to 0. There will be no variety of outcomes.

The statistical term for variety is <u>variance</u>. When we look at the statistical formula for <u>proportion variance,</u> we see that it captures the dynamic we have just seen.[1]

$$p \times (1 - p)$$

---

1 For those who are already acquainted with the general formula for variance shown below on the left-hand side, note that when x is a binomial variable with values 0 or 1, we can derive the specialized formula for the proportion variance from the general formula for variance. The symbol $\bar{x}$ stands for the arithmetic average (the mean) of all the x values.

$$\frac{1}{n}\sum_{i=1}^{n}(x_i - \bar{x})^2 = p \times (1 - p)$$

With a 50% chance of heads, $0.5 \times (1 - 0.5) = 0.5 \times 0.5 =$ 0.25 variance

With a 25% chance of heads, $0.25 \times (1 - 0.25) = 0.25 \times 0.75 =$ 0.1875 variance

With a 10% chance of heads, $0.1 \times (1 - 0.1) = 0.1 \times 0.9 =$ 0.09 variance

With a 0% chance of heads, $0.0 \times (1 - 0.0) = 0.0 \times 1.0 =$ 0.00 variance

As with sample sizes, it stands to reason that when the sample proportion histograms become less spread out, then the "verdict rule" intervals should become less spread out too.

When variance is *smaller*, our "verdict rule" can use *smaller* intervals.

It is also important to keep in mind that variance and uncertainty are intimately related, analogous to what we saw for sample size and uncertainty. The less variance there is, the less uncertainty there is about which outcomes will occur. Inversely, the more variance there is, the more uncertainty there is about which outcomes will occur.

Less variance → less uncertainty

As we'll see, the uncertainty due to variance $(p \times (1 - p))$ is incorporated into various statistical formulas.[2]

Together, sample size and sample variance
are key determinants of uncertainty and thus of interval width.

---

2 The square root of variance is called the standard deviation. We'll consider it in Part II.

7

Judging Opinion Splits I

We can put what we now know to use for public-opinion surveying. We'll ask people whether they agree or disagree with something. This agree or disagree item is a binomial variable. Our null hypothesis is that the community is evenly split in opinion, that is, 50% agree and 50% disagree overall. We'll gather our evidence by surveying 30 people. To avoid inadvertent <u>bias</u> when gathering sample opinions for our survey, everyone in the community must have an equal chance of being surveyed. Therefore, we'll select 30 people *at random*, giving us a <u>random sample</u> of 30 opinions. The 30 opinions serve as our evidence.

This surveying situation is analogous to coin flipping. Given our null hypothesis that the community is evenly split (coin is fair), by randomly selecting each person to survey (each flip) we expect a 0.5 probability of getting someone who agrees (head) and a 0.5 probability of getting a person who disagrees (tail). Surveying 30 random people in this context is analogous to flipping a coin 30 times. We want to see if we should reject the Null Hypothesis

Illuminating Statistical Analysis Using Scenarios and Simulations, First Edition.
Jeffrey E Kottemann.
© 2017 John Wiley & Sons, Inc. Published 2017 by John Wiley & Sons, Inc.

Table 7.1 Surveying outcomes.

Your verdict / Unknown truth	Community evenly split Do not reject null hypothesis	Community not evenly split Reject null hypothesis
Community is evenly split (null hypothesis)	Correct	Incorrect: type I error
Community is not evenly split	Incorrect: type II error	Correct

of an evenly split community (a fair coin). Table 7.1 lays out the possible surveying outcomes.

If our survey results give us 0% agree, we can be pretty sure that the overall community is not evenly split and we can reject the null hypothesis, because it is extremely unlikely to get a random sample with 0% from an evenly split community. Likewise, if 100% agree we can be pretty sure that the overall community is not evenly split and we can reject the null hypothesis, because it is also extremely unlikely to get a random sample with 100% agree from an evenly split community.

On the other hand, if the results give us 50% agree we can be sure that we do not have evidence supporting a rejection of the null hypothesis. It is fairly likely that we could get a random sample with 50% agree from an evenly split community. But what about if we get 40% or 30% or 20% agree? In between the obvious cases of 0%, 50, and 100% agree, statistical inference—like we did with coins—is very helpful.

Figure 7.1 is a histogram of simulation results showing the chances of various survey results that we can expect when we randomly sample from a community where the null hypothesis is true: evenly split between agree and disagree. It

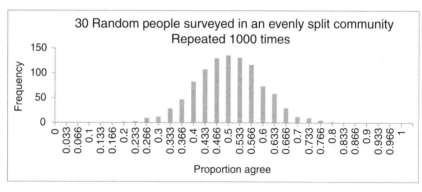

Figure 7.1

represents the results obtained by 1000 different independent surveyors who each randomly sampled 30 people from a community that is evenly split. The horizontal axis shows the various sample proportions of agreement. The vertical axis is the number of surveyors who got the various sample proportions.

If the null hypothesis is true, it is likely that our survey results will be in the vicinity of 0.5 (50%) agree, and it is very unlikely that our survey results will be in the vicinity of 0 (0%) agree or 1 (100%) agree. Let's set our alpha-level to be 0.05, corresponding to 95% confidence. Therefore, we want an interval containing 95% of the sample proportions.

Visually, it looks like the 95% interval is 0.333 (10 of 30) agree to 0.666 (20 of 30) agree. Approximately 950 of the sample proportions are inside and 50 are outside the 0.333–0.666 interval. That is 95% inside and 5% outside. So, based on visual inspection of the histogram, we could make the following rule: With a sample size of 30, if the sample proportion is outside the interval of 0.333–0.666, then reject the null hypothesis of an evenly split community, otherwise do not reject the null hypothesis.

As we did with the coin flipping example in Chapter 4, we could also determine the theoretical probabilities for all of the possible outcomes. But there is a better way. We can use the following "trick" formula *that takes into account the critical roles of variance and sample size.* (Do not worry about the details behind the "trick" for now. We'll uncover them gradually.)

We expect 95% of the sample proportions to be contained in the following interval:

$$\text{From } p - 1.96\sqrt{(p \times (1-p))/n} \text{ to } p + 1.96\sqrt{(p \times (1-p))/n}$$

In this case, p is the proportion of agree that we expect under the null hypothesis, which is 0.5. The proportion variance $(p \times (1-p))$ equals 0.25, and the sample size n is 30.

Calculations using the "trick" formula yield the following interval.

$$\text{From } 0.5 - 1.96\sqrt{(0.5 \times (1-0.5))/30} \text{ to } 0.5 + 1.96\sqrt{(0.5 \times (1-0.5))/30}$$
$$\text{From } 0.5 - 0.18 \text{ to } 0.5 + 0.18$$
$$\text{From } 0.32 \text{ to } 0.68$$

This interval is effectively the same as the interval we determined with the simulation: 0.32–0.68 contains the same outcomes as 0.333–0.666 (you can check the Figure 7.1 histogram to confirm this).

The "trick" formula works because the distribution of sample proportions becomes the extra-special bell-shape called the <u>normal distribution</u> when sample size is sufficiently large. The "trick" depends on this, and we'll go into much more detail about it as we go along.

We can use the verdict rule we just came up with for any binomial variable with a sample size of 30. For example, if we flip a coin 30 times, we can have an analogous rule. Verdict: If proportion of heads is outside the interval 0.32–0.68 (or equivalently 0.333–0.666), then reject the null hypothesis of a fair coin, otherwise do not reject the null hypothesis. With this rule, we can be 95% confident that we will not make a type I error. If instead, we flipped the coin 40 times or surveyed 40 people, we can recalculate the interval using 40 for n rather than 30 (or redo the simulation using a sample size of 40). Similarly, if we wanted to check whether the coin favored heads 60%, we can recalculate the interval using 0.6 for p rather than 0.5 (or redo the simulation using a probability of 0.6).

> The simulations are *recreations* of statistical phenomena.
> The formulas are *mathematical models* of statistical phenomena.

Here is a quick *statistical scenario*:

With an evenly split community, what is the chance that a surveyor would get a sample proportion of 0.2 agree with a random sample of size 30?

Answer: You can tell by the histogram that a sample proportion of 0.2 agree is extremely unlikely. The interval derived via the "trick" formula suggests likewise.

Below is a little more key terminology.

Normal distribution is the special bell-shaped distribution that allows us to use the "trick." We'll investigate this special distribution in more detail later, and we'll see it in action many, many times.

Confidence interval is the name for the verdict rule intervals we have been making. The interval in this chapter is a 95% confidence interval because we expect 95% of the possible outcomes to fall within the interval. We'll investigate other confidence intervals, especially the 99%, in more detail later, and we'll keep seeing confidence intervals in action many, many times.

Margin of error is the half-width of the 95% confidence interval, which is part of the formula we used: $1.96\sqrt{(p \times (1-p))/n}$. It equals 0.18 in this chapter's scenario. It can also be expressed as a percentage such as 18% rather than 0.18, and is often reported that way with poll results. Our 95% confidence interval for this chapter can be stated as $50\% \pm 18\%$ or 0.50 ± 0.18.

Standard error is this part of the formula: $\sqrt{(p \times (1-p))/n}$. It is the standard measure reflecting how spread out a normal sampling distribution is. We'll be seeing a lot of this too.

8

Amount of Evidence II

Statistical Scenario—Amount of Evidence II
What happens to our confidence interval as we increase our sample size?
What is a good sample size to use?

As I wrote at the end of Chapter 5, "the uncertainty due to sample size n is incorporated into various statistical formulas." Notice how the "trick" formula incorporates sample size.

$$\text{From } p - 1.96\sqrt{(p \times (1-p))/n} \text{ to } p + 1.96\sqrt{(p \times (1-p))/n}$$

Since n is in the denominator, the confidence interval narrows as the sample size n increases: When n increases, the standard error $\sqrt{(p \times (1-p))/n}$ decreases and so margin of error decreases. As the amount of evidence increases, uncertainty decreases and the confidence interval can narrow its focus.

Table 8.1 shows results when the margin of error $(1.96\sqrt{(p \times (1-p))/n})$ is calculated for various sample sizes with p equal to 0.5. Margin of error decreases as n increases. This is also what we saw with the sample proportion histograms using various sample sizes in Chapter 5.

Table 8.1 Margin of error for various sample sizes ($p = 0.5$).

Sample size n	30	60	90	120	150	180	210	240	270	300
Margin of error	0.179	0.127	0.103	0.089	0.080	0.073	0.068	0.063	0.060	0.057

Illuminating Statistical Analysis Using Scenarios and Simulations, First Edition.
Jeffrey E Kottemann.

Table 8.2 Margin of error for more sample sizes ($p = 0.5$).

Sample size n	100	500	1000	1500
Margin of error	9.8%	4.38%	3.10%	2.53%

What is a good sample size to use?

Table 8.2 shows margin of error with a wider range of sample sizes. Margin of error in this table is expressed as a percentage such as 9.8% (0.098) and the full 95% confidence interval would be expressed as $50\% \pm 9.8\%$ (0.50 ± 0.098). These margin of error values are for evenly split situations since the value of 0.5 (50%) was used for p in the calculations.

Using coin flipping as an example, the margin of error indicates that if you flip a fair coin 100 times, there is a 95% chance you will get between about 40.2% and 59.8% heads (about 40–60 actual heads). If you flip it 1000 times, there is a 95% chance you will get between 46.9% and 53.1% heads (469–531 actual heads).

Looking at the Table 8.2 entries, notice the *diminishing value of additional evidence.* You can see that adding an additional 500 observations to an existing sample of 500 (giving 1000 total sample size) decreases the margin of error by more than 1% from 4.38% to 3.10%. Adding an additional 500 observations to an existing sample of 1000 (giving 1500 total sample size) decreases the margin of error by less than 1% from 3.10% to 2.53%. The graph in Figure 8.1 illustrates the relationship between n and margin of error.

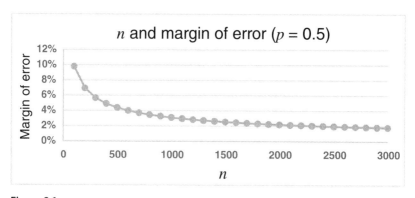

Figure 8.1

Professional public-polling firms often use sample sizes of 1000 or 1500 because, *for their situation,* those offer a good balance between the costs of gathering evidence and the benefits of precise estimates. Different situations will call for different balancing.

Statistical Scenario: Let's say a surveyor got a random sample with a sample proportion of 52%. Looking at the Figure 8.1 graph, approximately how large would the random sample have to be for the surveyor to reject the null hypothesis of an evenly split community? (The 52% needs to be outside the 95% confidence interval for 50% agree.)

Answer: We need to have n such that 50% plus the margin of error is less than 52%, so the margin of error must be less than 2%. Looking at the graph, it looks like n of 2500 is approximately where margin of error begins to be less than 2%.

9

Variance of Evidence II

Statistical Scenario—Variance of Evidence II

What happens to our confidence interval when the proportion of agree is not 50%?

What happens when the proportion gets close to the boundaries of 0 or 1?

As I wrote at the end of Chapter 6, "the uncertainty due to variance $(p \times (1 - p))$ is incorporated into various statistical formulas." Notice how the "trick" formula incorporates variance.

$$\text{From } p - 1.96\sqrt{(p \times (1 - p))/n} \text{ to } p + 1.96\sqrt{(p \times (1 - p))/n}$$

Recall that when p gets either smaller or larger than 0.5, then the proportion variance $(p \times (1 - p))$ gets smaller. In the "trick" formula, the term $(p \times (1 - p))$ is in the numerator, and so as p decreases or increases from 0.5, then the standard error $\sqrt{(p \times (1 - p))/n}$ decreases and so margin of error decreases. As variance decreases, uncertainty decreases and the confidence interval can narrow its focus.

You can see this in Table 9.1. The Table shows results when the margin of error $(1.96\sqrt{(p \times (1 - p))/n})$ is calculated for various values of p with sample size of 30. The margin of error decreases as p moves from 0.5. This is also what we saw with the sample proportion histograms using various values for p and variances $(p \times (1 - p))$ in Chapter 6.

Table 9.1 Margin of error for various proportions ($n = 30$).

Proportion p	0	0.1	0.2	0.3	0.4	0.5	0.6	0.7	0.8	0.9	1
Margin of error	0.000	0.107	0.143	0.164	0.175	0.179	0.175	0.164	0.143	0.107	0.000

Illuminating Statistical Analysis Using Scenarios and Simulations, First Edition.
Jeffrey E Kottemann.
© 2017 John Wiley & Sons, Inc. Published 2017 by John Wiley & Sons, Inc.

What happens when the proportion gets close to the boundaries of 0 or 1?

Looking at the Table 9.1 entries, notice that the margin of error values in the vicinity of p equal to 0.5 are fairly similar and that the margin of error shrinks more rapidly as p approaches 0 or 1. The graph in Figure 9.1 illustrates the relationship between p and margin of error.

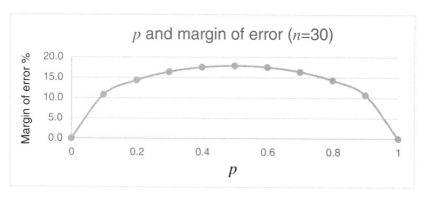

Figure 9.1

Although the distribution of sample proportions condenses as p approaches 0 or 1—and so the margin of error shrinks—it may not condense enough to avoid bumping into those boundaries. When p becomes too close to 0 or 1, the distribution deforms from its characteristic normal bell shape. You can see this in the simulation histogram shown in Figure 9.2 for p of 0.9 with n of 30. Such deformations undermine the "trick" formula because the "trick" assumes that a normal bell shape exists—the upper-bound of the 95% confidence interval calculated using the formula is $0.9 + 0.107 = 1.007$, which is greater than 1.0 and is infeasible.

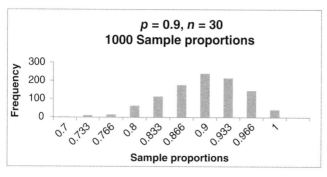

Figure 9.2

Larger sample sizes help alleviate the problem since they further condense the distribution: Compare the $n = 30$ simulation histogram in Figure 9.2 with the $n = 100$ simulation histogram in Figure 9.3.

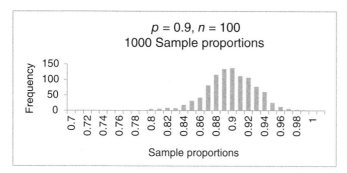

Figure 9.3

A good rule of thumb for the *minimum sample size* that is needed for the "trick" formula to work and which takes into account the relationship between p and n is to make sure n is large enough so that $n \times p \geq 10$ and $n \times (1 - p) \geq 10$. Using this rule of thumb for p of 0.9, we can determine that n of 30 is too small because $30 \times (1 - 0.9)$ only equals 3, and that n of 100 is just big enough because $100 \times (1 - 0.9)$ just equals 10. With p of 0.5, the minimum sample size is 20.

In situations where p is too close to 0 or 1, there are other (more complicated) formulaic methods that can be used, but we will not cover them here. Simulation can always be used to shed light on things.

10

Judging Opinion Splits II

Rather than checking whether a community as a whole might be evenly split, let's figure out how to use our sample proportion statistic and confidence interval to determine what the actual overall community opinion split might actually be.

Statistical Scenario—Survey II

Pollster #1 randomly surveyed 100 community members about a proposed new local policy and found that 40 of the 100 agreed with the proposed new policy. The 0.40 sample proportion is the evidence.

Given the sample proportion of 0.40, draw two lines to make an interval that you think should very likely contain the true, but unknown, overall community proportion (the proportion you would get if you surveyed absolutely everyone).

    0   0.1   0.2   0.3   0.4   0.5   0.6   0.7   0.8   0.9   1.0
                            Proportion of agree

Also, in the same community using the same survey and sample size, pollster #2 got a sample proportion is 0.35 and pollster #3 got 0.30. What are their interval lines?

Now, let's say that the proportion of agree for the entire community is actually 0.40. Only omniscient beings know this. We'll use the 0.40 community proportion of agree for a survey sampling simulation. The histogram in Figure 10.1 shows what happened when 1000 independent surveyors each surveyed 100 random community members. The histogram shows the distribution of the 1000 sample proportions.

Illuminating Statistical Analysis Using Scenarios and Simulations, First Edition.
Jeffrey E Kottemann.
© 2017 John Wiley & Sons, Inc. Published 2017 by John Wiley & Sons, Inc.

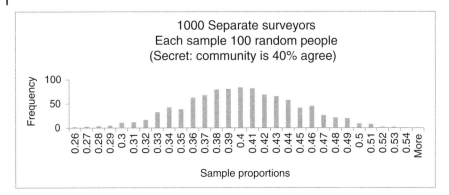

Figure 10.1

Based on visual inspection, notice that a great majority of the 1000 surveyors' sample proportions are in the interval 0.31–0.49. Approximately 950 of the 1000 surveyors' sample proportions are contained in the interval 0.31–0.49, suggesting that 0.31–0.49 is the 95% confidence interval surrounding the true community proportion of 0.40. The "trick" formula will yield effectively the same 95% confidence interval. Try it and see.

> We expect the 95% confidence interval around the true community proportion to contain 95% of all sample proportions obtained by random sampling.

But how does each of the individual surveyors view the situation? They don't know the true community proportion, and they only know their own individual survey result.

First, let's relook at the "trick" formula for calculating 95% confidence intervals.

$$\text{From } p - 1.96\sqrt{(p \times (1 - p))/n} \text{ to } p + 1.96\sqrt{(p \times (1 - p))/n}$$

Each of the 1000 surveyors would calculate their individual interval using their sample proportion for p, and we expect that 95% of the surveyors' intervals will contain the true community proportion of 0.4.

> The 1000 surveyors calculate their individual 95% confidence intervals. About 950 of them will have an interval containing the true community proportion. About 50 of them will not.

Let's look at the 95% confidence intervals for the three surveyors of the scenario: The first who got a sample proportion of 0.40, the second who got a sample proportion of 0.35, and the third who got a sample proportion of 0.30.

Statistical Scenario—Survey II Result #1

Using the formula with a sample proportion of 0.40 and a sample size of 100, the 95% confidence interval is 0.304–0.496.

| 0 | 0.1 | 0.2 | 0.3 | 0.4 | 0.5 | 0.6 | 0.7 | 0.8 | 0.9 | 1.0 |
Proportion of agree

Only the omniscient beings know that this 95% confidence interval contains the true community proportion of 0.4.

Statistical Scenario—Survey II Result #2

Using the formula with a sample proportion of 0.35 and a sample size of 100, the 95% confidence interval is 0.257–0.443.

| 0 | 0.1 | 0.2 | 0.3 | 0.4 | 0.5 | 0.6 | 0.7 | 0.8 | 0.9 | 1.0 |
Proportion of agree

Only the omniscient beings know that this 95% confidence interval contains the true community proportion of 0.4.

Statistical Scenario—Survey II Result #3

Using the formula with a sample proportion of 0.30 and a sample size of 100, the 95% confidence interval is 0.210–0.390.

| 0 | 0.1 | 0.2 | 0.3 | 0.4 | 0.5 | 0.6 | 0.7 | 0.8 | 0.9 | 1.0 |
Proportion of agree

Only the omniscient beings know that this 95% confidence interval does not contain the true community proportion of 0.4.

Only the omniscient beings know that this survey is one of the expected 5% type I error victims.

Summary: (1) We expect the 95% confidence interval around the true community proportion to contain 95% of all sample proportions obtained by random sampling. (2) We expect 95% of all the 95% confidence intervals around random sample proportions to contain the true community proportion.

Table 10.1 shows what various surveyors will get. Notice that the expected 950 surveyors with sample proportions in the interval 0.31–0.49 also have 95% confidence intervals that contain the true community proportion of 0.40. The expected 50 surveyors with sample proportions outside the interval 0.31–0.49 do not.

Table 10.1 Confidence intervals for various surveyors.

Surveyor's sample proportion	95% CI low	95% CI high	Surveyor's sample proportion	95% CI low	95% CI high	Surveyor's sample proportion	95% CI low	95% CI high
0.27	0.183	0.357	0.36	0.266	0.454	0.45	0.352	0.548
0.28	0.192	0.368	0.37	0.275	0.465	0.46	0.362	0.558
0.29	0.201	0.379	0.38	0.285	0.475	0.47	0.372	0.568
0.30	0.210	0.390	0.39	0.294	0.486	0.48	0.382	0.578
0.31	0.219	0.401	0.40	0.304	0.496	0.49	0.392	0.588
0.32	0.229	0.411	0.41	0.314	0.506	0.50	0.402	0.598
0.33	0.238	0.422	0.42	0.323	0.517	0.51	0.412	0.608
0.34	0.247	0.433	0.43	0.333	0.527	0.52	0.422	0.618
0.35	0.257	0.443	0.44	0.343	0.537	0.53	0.432	0.628

Here is a quick *Statistical Scenario*. Suppose a stamping plant that makes coins was malfunctioning and produced unfair coins. Unbeknownst to anyone, these unfair coins favored tails, and the chance of coming up heads is only 0.4. Now say 1000 people flip these coins 100 times each, while counting and then determining the proportion of heads. What would their 1000 results be like? What would an analysis of a single coin and its 100 flips be like?

Answer: Just like the survey example above. Just replace the words "agree" and "disagree" with the words "heads" and "tails."

Below are two more key terms.

(From now on, I will underline and explain new terminology as it comes up.)

Population is the group(s) you are investigating and want to generalize your results to. In the current example, it is the entire community. If, instead, you wanted to investigate only community females, you would say that you will sample from the population of community females. If you wanted to investigate registered voters across the United States, you would say that you will sample from the population of registered voters in the United States. You sample *randomly* from the population in order to get unbiased sample statistics that are used as estimates for the true population statistics. (There are other sampling strategies, but we'll stick with random sampling for our examples.)

Population statistic is the unknown, true overall value that you are trying to estimate using sample statistics. At the beginning of this chapter, we said the true population (community) proportion was 0.40, but that only imaginary omniscient beings can know such things. The earthly surveyors used surveys to get sample proportions to use as estimates of the population proportion.

11

It Has Been the Normal Distribution All Along

I remarked in an earlier chapter that this "trick" formula

$$p - 1.96\sqrt{(p \times (1 - p))/n} \text{ to } p + 1.96\sqrt{(p \times (1 - p))/n}$$

works because the shape of the sampling distribution histograms we have been seeing are expected to become the extra-special bell-shape called the <u>normal distribution</u> when our sample size is large enough.[1]

The 1.96 in the formula is there instead of some other number because the distribution is a normal distribution. And with a normal distribution of binomial proportions, we expect 95% of its contents to lie within the interval of $p \pm 1.96$ times $\sqrt{(p \times (1 - p))/n}$. This is a fact about the normal distribution. It is a fact much like pi (π) is a fact about the relationship between a circle's diameter and circumference ($c = \pi \times d$).

Nature often does not give us nice round numbers. Pi is 3.14159 . . . with seemingly random digits going on and on. For simplicity we often round it and write 3.14 ($c = 3.14 \times d$). The figure 1.96 is rounded too. It is closer to 1.9599639845, but 1.9599639845 is *so close* to 1.96 that we usually just write 1.96.

As we have seen, this $\sqrt{(p \times (1 - p))/n}$ has a name: <u>standard error</u>. Its full name is <u>standard error of a sample proportion</u>. It is labeled "standard" because it serves as a standard unit. And it is labeled "error" because we don't expect our sample proportions to be exactly equal to the true population proportions. As we have seen, the magnitude of the standard error of a sample proportion reflects how spread out the normal distribution of sample proportions is. We'll refer to standard error a lot.

Since 0.05 and 0.01 thresholds (alpha-levels), corresponding to 95% and 99% confidence-levels, respectively, are often used in practice, it behooves you to memorize two things related to the normal distribution:

---

[1] The mathematical proof of this is the de Moivre–Laplace theorem, which is a special case of the central limit theorem. More on this later.

Illuminating Statistical Analysis Using Scenarios and Simulations, First Edition.
Jeffrey E Kottemann.
© 2017 John Wiley & Sons, Inc. Published 2017 by John Wiley & Sons, Inc.

> 1.96 Standard errors is the 95% confidence-level. (Alpha-level of 0.05)
> 2.58 Standard errors is the 99% confidence-level. (Alpha-level of 0.01)

Below is the *Statistical Scenario—Survey II Result #3* from the previous chapter, now with the 99% confidence interval added. It is calculated using the "trick" formula, again with 0.3 for the sample proportion *p* and again with 100 for the sample size *n*, but now with 2.58 in place of the 1.96.

Statistical Scenario—Survey II Result #3 Revisited

Using the formula with a sample proportion of 0.30 and a sample size of 100, the 95% confidence interval is 0.210–0.390, and the 99% confidence interval is 0.182–0.418.

Only the omniscient beings know that, whereas the 95% confidence interval does not contain the true proportion, the 99% confidence interval does.

The 99% confidence interval casts a wider net in order to be more confident of catching the true population proportion.

> The 1000 surveyors determine their individual 99% confidence intervals. About 990 of them will have an interval containing the true community proportion. About 10 of them will not.

A Note on Stricter Thresholds for Type I Error

While these two thresholds are seen most often, certain fields use much stricter thresholds. In particle physics, 5.0 standard errors is used. This corresponds to an alpha-level of approximately 0.0000003. Particle physicists want to be extraordinarily confident before they declare the existence of a new subatomic particle. Using this alpha-level for coin flipping, we would insist that a coin come up heads at least 75 times in 100 flips before declaring it unfair.

Biologists conducting what are called genome-wide association studies also use very strict thresholds. When these researchers test hundreds of thousands or even millions of genome locations, Type I error would be far too frequent if they used an alpha-level of 0.05 or 0.01. Instead, they use an alpha-level of 0.00000005. Think about it this way: If you flipped 100,000 separate fair coins and used an alpha-level of 0.05, you would expect to erroneously declare about 5000 of the fair coins to be unfair. If you used an alpha-level of 0.00000005, you would not expect to declare any of the fair coins to be unfair (although on rare occasion it would happen).

12

Judging Opinion Split Differences

Statistical Scenario—Surveying for Differences

A survey was conducted in the Flowing Wells (FW) community in which 100 random members were asked their opinion of a new state sales tax policy. The same survey was conducted in the Artesian Wells (AW) community asking 100 random members their opinions. Which of the *sample proportion differences* for the two communities shown in Table 12.1 might reasonably lead you to believe that there is a *difference in the overall opinion split between the two community populations*? Just use your intuition.

Null hypothesis:
No difference between the FW and the AW communities' population proportions. That is, the difference between the FW and the AW population proportions equals zero.

Table 12.1 Various sample proportion differences.

Sample proportion differences between the two communities (FW sample proportion minus AW sample proportion)
0.05 (e.g., 0.55–0.50)
0.10
0.15
0.20
0.25

Illuminating Statistical Analysis Using Scenarios and Simulations, First Edition.
Jeffrey E Kottemann.
© 2017 John Wiley & Sons, Inc. Published 2017 by John Wiley & Sons, Inc.

Rephrasing: How likely is it to get each of the sample proportion differences if in fact the difference between the FW and the AW population proportions equals zero?

The null hypothesis is that there is no difference between the FW and the AW communities' population proportions, so let's look at simulation results that assume this is true. We'll assume that both Flowing Wells and Artesian Wells have an overall community population opinion with 50% (0.50) agree. If we randomly surveyed 100 people in each of the two identical communities, how likely is it that the sample proportion differences shown in the table could arise just by chance, due to the randomness inherent in the sampling?

Rather than our customary 1000 surveyors, we'll employ 10,000 to help ensure that we get a dense, relatively smooth histogram. The following simulation is equivalent to 10,000 surveyors independently surveying 100 random people in Flowing Wells and 100 random people in Artesian Wells. Unbeknownst to anyone, both communities have 50% who agree and 50% who disagree. The surveyors calculate each of their two sample proportions and the difference between the two. All 10,000 surveyors then report their differences to someone who counts how many of the differences fall into various "bins" to form a histogram. Figure 12.1 is the end product.

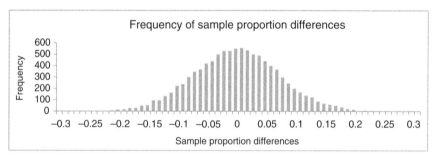

Figure 12.1

How likely are random sample proportion differences of 0.05, 0.10, 0.15, 0.20, or 0.25 to occur by chance if the two community populations in fact have the same average opinion? The sample proportion differences of 0.05 and 0.10 appear relatively likely to occur by chance; they are clearly inside the main body of the histogram. On the other hand, 0.15, 0.20, and 0.25 are out in the right "tail" of the histogram. (The extreme ends of a distribution—left and right—are called "tails"; it has nothing to do with coins.) Table 12.2 summarizes these conclusions.

Table 12.2 Surveying for differences answer #1.

Proportion difference	Does the sample proportion difference provide strong evidence of a difference between the population proportions?
0.05	No, it is inside the main body of the histogram.
0.10	No, it is inside the main body of the histogram.
0.15	Ok, it is outside the main body of the histogram; it is in the tail.
0.20	Yes, it is definitely out in the tail.
0.25	Yes, it is definitely far out in the tail.

Let's actually get a rough estimate for how unlikely a difference of 0.15 or more is: (1) we locate the 0.15 difference on the histogram and (2) determine how many sample proportions lie that far out or more. Eyeballing, it looks like a total of about 200 out of 10,000, or 2%, are in the 0.15 bar plus the other bars out to the right. So, the approximate probability of being that far out (or more) is 0.02.

Whether the difference we are considering is positive (0.15) or negative (−0.15) simply depends on whether we have calculated $FW - AW$ or $AW - FW$. So, we typically consider both 0.15 and −0.15 and include the "mirror-image" by adding in the probability of −0.15 or less too. This doubles the approximate probability to 0.04. The 0.04 is a two-tailed probability because it involves both tails of the histogram's distribution. (More on this later.)

The other sample proportion differences can be analyzed this way as well, but it is obvious that the sample proportion differences 0.05 and 0.10 have fairly large probabilities, and that the sample proportion differences 0.20 and 0.25 have very small probabilities.

Probabilities such as the 0.04 we just came up with are typically called *p*-values for short. They are used in deciding whether to reject the null hypothesis.

When the *p*-value is less than the alpha-level, reject the null hypothesis.

When the *p*-value is less than the alpha-level, call the result statistically significant.

Let's say our alpha-level is set at the common threshold of 0.05. Our null hypothesis is that there is no difference between the FW and AW communities' population proportions. The sample proportion difference of 0.15 corresponds to a two-tail *p*-value of approximately 0.04. Since the *p*-value of 0.04 is less than the alpha-level of 0.05, we reject the null hypothesis and say that the 0.15 difference is statistically significant.

There is only about a 4% chance that we would get a sample proportion difference this extreme when two communities have the same agree-or-disagree opinion split. In other words, the sample proportion difference of 0.15 is not indicative of two identical communities. Therefore, we conclude that the communities do not have the same agree-or-disagree opinion split. In everyday language, the surveyor might say "I probably would not have gotten sample proportions that are this different if the two communities have the same average opinion, so I guess they don't have the same average opinion."

13

Rescaling to Standard Errors

You can see from the previous histogram that the distribution of sample proportion *differences* approximates a normal distribution, and indeed they do in general. So, there is a corresponding "trick" formula that can be used, which we'll look at next, followed by determining *p*-values using standard errors and the standard normal distribution.

The formulaic method we'll explore in the next few chapters serves the same purpose as the "simulation results histogram eyeballing" method we just used. The formulaic method calculates the number of standard errors for the difference between two sample proportions, and then determines the *p*-value for that number of standard errors of the standard normal distribution. The below formula does the first step by *rescaling* a proportion difference into the unit of standard errors. It is simply the proportion difference divided by the appropriate standard error, yielding the number of standard errors, #SEs, for the sample proportion difference. This is called <u>standardizing</u>.

$$\frac{p_1 - p_2}{\sqrt{\dfrac{p(1-p)}{n_1} + \dfrac{p(1-p)}{n_2}}} = \#\text{SEs}$$

p_1 is the first sample proportion, p_2 is the second sample proportion, p is the average of the two (weighted average if you have unequal sample sizes), n_1 is the size of the first sample, and n_2 is the size of the second sample. The numerator is the difference between our sample proportions, and the denominator is the <u>standard error of the difference between two sample proportions</u>. As with the formula for a single proportion, the standard error formula takes into account the sample variances and sample sizes.

Table 13.1 shows the calculated results of the formula for various values of p_1 and p_2 all with sample sizes of 100. The cases from the *statistical scenario* are highlighted. Whether the result of the #SEs formula comes out positive or negative is simply a matter of which sample proportion is subtracted from which. This is inconsequential because we'll be doing two-tailed analyses.

Illuminating Statistical Analysis Using Scenarios and Simulations, First Edition.
Jeffrey E Kottemann.
© 2017 John Wiley & Sons, Inc. Published 2017 by John Wiley & Sons, Inc.

Table 13.1 Standard error results for various sample proportion differences.

For cases when both sample sizes = 100					
p_1	p_2	Average (p)	Difference	Standard error	#SEs
0.50	0.50	0.500	0.00	0.07071	0.00
0.55	0.50	0.525	0.05	0.07062	0.71
0.60	0.50	0.550	0.10	0.07036	1.42
0.65	0.50	0.575	0.15	0.06991	2.15
0.70	0.50	0.600	0.20	0.06928	2.89
0.75	0.50	0.625	0.25	0.06847	3.65
0.80	0.50	0.650	0.30	0.06745	4.45
0.85	0.50	0.675	0.35	0.06624	5.28

Notice how the two differences we previously said *were not* extreme enough to reject the null hypothesis because they are in the body of the histogram (0.05 & 0.10) have the number of standard errors (#SEs) less than 1.96 and therefore have more than a 5% chance of type I error. And the three differences we said *were* extreme enough to reject the null hypothesis because they are in the tails of the histogram (0.15, 0.20, and 0.25) have #SEs greater than 1.96 and therefore have less than a 5% chance of type I error. Moreover, the #SEs of the differences 0.20 and 0.25 are greater than 2.58, so there is less than a 1% chance of type I error for those two cases. Table 13.2 summarizes these conclusions.

Table 13.2 Surveying for differences answer #2.

Proportion difference	Does the sample proportion difference provide strong evidence of a difference between the population proportions?
0.05 (or −0.05)	No, #SE = 0.71 < 1.96
0.10 (or −0.10)	No, #SE = 1.42 < 1.96
0.15 (or −0.15)	Ok, #SE = 2.15 > 1.96
0.20 (or −0.20)	Yes, #SE = 2.89 > 2.58
0.25 (or −0.25)	Yes, #SE = 3.65 > 2.58

14

The Standardized Normal Distribution Histogram

As we have seen, random sampling scenarios for sample proportions and sample proportion differences form normal distributions. And, we can rescale all of them to a common scale: The #SEs scale. Let's put our most recent histogram of sample proportion differences into the standardized form. Figure 14.1 is the original histogram.

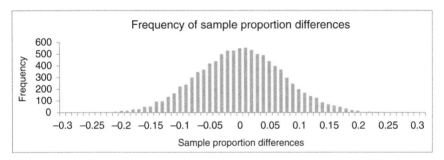

Figure 14.1

In the standardized histogram of Figure 14.2, the horizontal axis is rescaled by using #SEs instead of proportion differences. Since it uses #SEs, it is a general-purpose histogram, as we'll see. The vertical axis is also rescaled by using relative frequency instead of frequency. Since it uses relative frequency, the sum of all the bar heights equals 1. Recall that relative frequencies can serve as probability estimates.

Illuminating Statistical Analysis Using Scenarios and Simulations, First Edition.
Jeffrey E Kottemann.

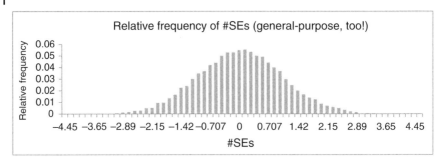

Figure 14.2

This standardized histogram of results approximates the *general-purpose standardized normal distribution*, which dovetails with the formulaic methods. Since the #SEs formula takes into account the sample proportions, their variances, and their sample sizes, we can use #SEs and the standardized normal distribution for any scenarios involving any sample proportions (except those too close to 0 or 1) and any sample sizes (except when they are too small). In short, we can calculate standard error and #SEs using the formula and then use the general-purpose histogram to determine relative frequencies.

> With sampling distribution histograms derived via simulation, the distributions expand and contract based on sample size and variance (Chapters 5 and 6).
>
> With standard error derived via formula, the standard error itself expands and contracts based on sample size and variance (Chapters 8 and 9), and because of that, we can use #SEs to reference the general-purpose standardized normal distribution.

In case you are not convinced, let's look at the sample proportion difference of 0.15 with sample sizes of 100 and then with sample sizes of 200.

First, let's review the evaluation we did previously for sample sizes of 100. Looking at the sample proportion frequency histogram with sample sizes of 100 (reproduced in Figure 14.1), we estimated the two-tail *p*-value to be about 0.04. Using the #SEs formula, we calculated standard error to be 0.06991 and #SEs to be $0.15/0.06991 = 2.15$ (see the table in the previous chapter). Looking at the general-purpose standardized normal histogram of Figure 14.2, we see that the two-tail *p*-value for #SEs of 2.15 yields the same *p*-value estimate of 0.04.

Now, for comparison, look at the Figure 14.3 sample proportion difference frequency histogram for the larger sample sizes of 200. *The histogram contracts* relative to the histogram for sample sizes of 100—no surprise there. The sample proportion difference of 0.15 is now extremely unlikely; the *p*-value is obviously less than 0.01.

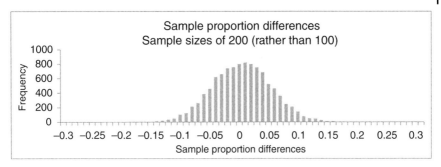

Figure 14.3

Using the #SEs formula with sample sizes of 200, *the Standard Error contracts to 0.04943* (relative to 0.06991 with sample sizes of 100). So, #SEs for the sample proportion difference of 0.15 increases to $0.15/0.04943 = 3.03$. Looking at the general-purpose standardized normal histogram of Figure 14.2, we can see that #SEs of 3.03 is extremely unlikely; the *p*-value is obviously less than 0.01. Same result.

15

The z-Distribution

If we use a continuous line to connect the tops of each bar of the general-purpose standardized normal distribution histogram from the previous chapter, we get a shape that looks like Figure 15.1.

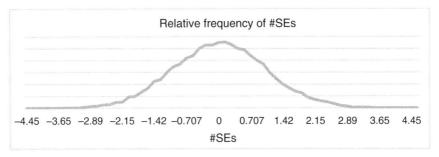

Figure 15.1

If we repeat the simulation many more times and use narrower bars, we get the smoother curve as shown in Figure 15.2.

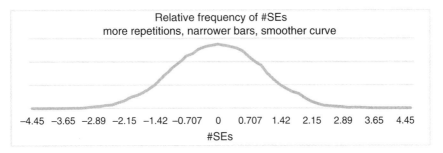

Figure 15.2

Illuminating Statistical Analysis Using Scenarios and Simulations, First Edition.
Jeffrey E Kottemann.
© 2017 John Wiley & Sons, Inc. Published 2017 by John Wiley & Sons, Inc.

If we could repeat the simulation an infinite number of times and could make the histogram bars infinitely narrow, we would get a general-purpose normal distribution curve that is perfectly smooth and continuous. It is called the z-distribution and is shown in Figure 15.3. It is a very important probability distribution.

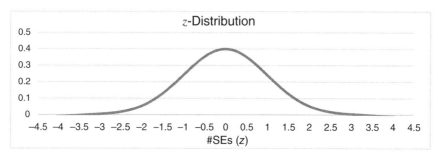

Figure 15.3

The z-distribution is a probability distribution that is normal, standardized, and continuous. Since it is a probability distribution, the entire area under the curve is 1. And, since the bars are infinitely narrow, probabilities for specific values—such the probability of #SEs exactly equaling 2—are zero. We always need to refer to the probabilities of value ranges—such as the probability of #SEs $\geq$ 2.

Table 15.1 shows some #SEs values and their corresponding two-tailed p-values. When we are using #SEs to get p-values from the z-distribution, the #SEs also go by the designation z. In other words, "#SEs" and "z" are synonymous when we are using #SEs to reference the z-distribution. It is a labeling convention. The two-tailed p-values are the areas of the z-distribution from #SEs (z) to positive infinity plus negative #SEs (−z) to negative infinity. For example, with #SEs (z) of 1, the p-value is the area to the left of −1 and to the right of +1. Rounded to the nearest 1000ths place, this equals 0.317.

Table 15.1 #SEs (z) and two-tail p-values using the z-distribution.

#SEs (z)	1	1.96	2	2.5	2.58	3
p-value (two-tail)	0.317	0.050	0.046	0.012	0.010	0.003

Mathematically, the z-distribution is defined by a hard-won formula.[1] To determine p-values with continuous distributions, there are no individual

---

1 The z-distribution is defined by $\frac{1}{\sqrt{2\pi}}e^{\left(-\frac{x^2}{2}\right)}$, where x is #SEs for our purposes. This formula was used to make the z-distribution curve illustration, using a series of values from −4.5 to +4.5. Appendix B gives the spreadsheet version of this formula along with instructions for "drawing" the curve. You can find more information here: http://itl.nist.gov/div898/handbook/eda/section3/eda3661.htm

histogram bars to measure and add up to get the area out in the tails because the bars are infinitely narrow. Integral calculus and numerical integration are used to determine these *p*-value areas. *P*-value calculators and statistical analysis software packages do such *z*-tests for you.

If you do an online search for "*p*-value calculator" you will find an assortment of free online calculators to choose from. *P*-value calculators will produce *p*-values for entered values of *z* using the *z*-distribution.[2] To use a *p*-value calculator for the FW versus AW proportion differences scenario, simply enter a value for *z* (#SEs) and the calculator will give the *p*-value. Table 15.2 is an expanded version of Table 13.1 showing the two-tailed *p*-values for the various cases we have been assessing. For a given *z* (#SEs) value in the table, the *p*-value is telling you how likely it is to be that far out or more in the *z*-distribution's tail on both sides (two-tailed). For example, the proportion difference of 0.15 rescales to *z* (#SEs) of 2.15 that corresponds to a two-tail *p*-value of 0.032. Recall that we estimated the *p*-value to be about 0.04 by informally eyeballing the sample proportion differences frequency histogram constructed via simulation back in Chapter 12.

Table 15.2 Two-tail *p*-value results for various sample proportion differences.

For cases when both sample sizes = 100						
p_1	p_2	Average (*p*)	Difference	Standard error	*z* (#SEs)	Two-tailed *p*-value
0.50	0.50	0.500	0.00	0.07071	0.00	1.00000
0.55	0.50	0.525	0.05	0.07062	0.71	0.47895
0.60	0.50	0.550	0.10	0.07036	1.42	0.15522
0.65	0.50	0.575	0.15	0.06991	2.15	0.03191
0.70	0.50	0.600	0.20	0.06928	2.89	0.00389
0.75	0.50	0.625	0.25	0.06847	3.65	0.00026
0.80	0.50	0.650	0.30	0.06745	4.45	0.00001
0.85	0.50	0.675	0.35	0.06624	5.28	0.00000

---

2 They will also determine *p*-values for the other types of probability distributions we'll see later on.

Notice that when z (#SEs) is zero, the p-value is the entire probability distribution (both halves), with area of 1.0. As z (#SEs) moves further and further from zero the p-value decreases, quickly at first and then more slowly because of the bell-shape.

Figure 15.4 shows approximately where each of our proportion difference scenarios map onto the standardized normal z-distribution. The p-values given in Table 15.2 are two-tailed p-values, so you need to imagine the mirror-image on the left side of the z-distribution with those mirror-image areas added in too.

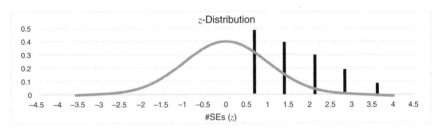

Figure 15.4

Now we can answer the questions posed earlier using the results of z-tests as shown in Table 15.3.

Table 15.3   Surveying for differences answer #3.

Proportion difference	Does the sample proportion difference provide strong evidence of a difference between the population proportions?
0.05 (or −0.05)	No, z (#SE) = 0.71; two-tail p-value = 0.48
0.10 (or −0.10)	No, z (#SE) = 1.42; two-tail p-value = 0.15
0.15 (or −0.15)	Ok, z (#SE) = 2.15; two-tail p-value = 0.03
0.20 (or −0.20)	Yes, z (#SE) = 2.89; two-tail p-value <0.01
0.25 (or −0.25)	Yes, z (#SE) = 3.65; two-tail p-value <0.001

> The z-distribution is a continuous, standardized, normal, probability distribution. It is an abstract representation of a wide range of statistical phenomena.

In addition to analyzing differences between two sample proportions, we can also reframe scenarios involving a *single* sample proportion into a proportion difference suitable for the z-distribution. Let's do a z-test for the scenario in

Chapter 10 where pollster #1 found that 40 of 100 agreed with the proposed new policy. We can assess whether the population proportion might be evenly split (0.5 agree) by rescaling the difference $0.4 - 0.5$ to standard error. Recall that the standard error for a single sample proportion is $\sqrt{(p \times (1 - p))/n}$, so we have

$$\#SEs\ (z) = (0.4 - 0.5)/\sqrt{(0.4 \times (1 - 0.4))/100} = 2.04$$

The sample proportion of 0.4 is -2.04 standard errors away from a population proportion of 0.5. The two-tail p-value for -2.04 is 0.04. So, at the 95% confidence-level (0.05 alpha) we would reject the null hypothesis of an evenly split population. This result is consistent with the 95% confidence interval for 0.4 that we calculated in Chapter 10: the interval 0.304–0.496 does not contain the proportion 0.5, although it is very close just as the p-value of 0.04 is very close to the 0.05 alpha-level.

> The formulaic approaches we have seen in Part I hinge on the fact that proportions and proportion differences are normally distributed. Therefore, we can convert them to the standard error scale and determine p-values using the standard normal z-distribution.

> The simulations are *recreations* of statistical phenomena. The formulas and continuous probability distributions are *mathematical models* of statistical phenomena.

We'll continue to analyze statistical scenarios via simulation, and bring in formulas, continuous probability distributions, and p-value calculators when the time is right.

16

Brief on Two-Tail Versus One-Tail

Statistical Scenario—Two-Tail, One-Tail

Say you want to see if a coin is fair or not, so you flip it *100 times* counting the number of heads.

Use an alpha-level of 0.05 (95% confidence level).

Where do you draw the two lines for declaring the coin to be *not fair*? (*Two lines*, one toward the left and the other line toward the right.)

Where do you draw the one line for declaring the coin *favors tails*? (*One line* toward the left.)

Where do you draw the one line for declaring the coin *favors heads*? (*One line* toward the right.)

0.38 0.40 0.42 0.44 0.46 0.48 0.50 0.52 0.54 0.56 0.58 0.60 0.62

Proportion of heads

The first question asks for two lines, as we have done before. It is called a two-tailed test for this reason. Recall that "tail" in this context means the far left and far right of a histogram's distribution, which are called the tails of the distribution.

For this chapter, a simulation of 10,000 people each flipping fair coins 100 times is used. In the histogram of Figure 16.1, the lines are drawn so that about 5% is divided equally between the two tails: about 2.5% of the simulation outcomes are in the left tail and about 2.5% are in the right tail. A tested coin outside the interval 0.41–0.59 will be declared not fair (reject the two-tailed null hypothesis).

Illuminating Statistical Analysis Using Scenarios and Simulations, First Edition.
Jeffrey E Kottemann.
© 2017 John Wiley & Sons, Inc. Published 2017 by John Wiley & Sons, Inc.

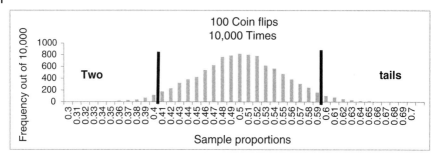

Figure 16.1

Next: Where do you draw the one line for declaring the coin *favors tails*? (*One line* toward the left.)

This asks for a <u>one-tailed test</u>. More specifically, it asks for a left-tail test (also called a lower-tail test). The null hypothesis is that the coin does not favor tails.

In this case, shown in Figure 16.2, the original line on the left is moved to the right so that 5% of the simulation outcomes are in the left tail. The original line on the right is eliminated. A tested coin with a proportion of heads less than 0.43 will be declared to favor tails.

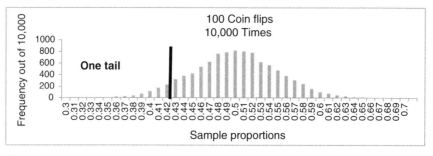

Figure 16.2

Next: Where do you draw the one line for declaring the coin *favors heads*? (*One line* toward the right.)

And this asks for a one-tailed test. More specifically, it asks for a right-tail test (also called an upper-tail test). The null hypothesis is that the coin does not favor heads.

In this case, shown in Figure 16.3, the original line on the right is moved left so that 5% of the simulation outcomes are in the right tail. The original line on the left is eliminated. A tested coin with a proportion of heads greater than 0.57 will be declared to favor heads.

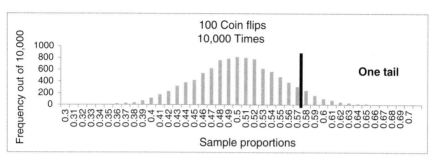

Figure 16.3

One-tailed hypotheses and statistical tests are used when you can convincing argue *beforehand* that a phenomenon will be in a certain *direction*. However, even though you may think you are investigating a directional phenomenon, a nondirectional hypothesis and its corresponding two-tailed test is usually considered more objective. In some situations, one-tailed tests may even seem a bit underhanded. Consider a coin to be objectively tested for fairness that produced a sample proportion of 0.58 heads. And, let's say that the tester has a vested interest and wants to declare the coin unfair. Using the two-tail test (Figure 16.1), however, the tester *cannot reject* the two-tailed null hypothesis because 0.58 *is not above* the threshold of 0.59. But, by changing to the right-tail test (Figure 16.3) the tester *can reject* the one-tailed null hypothesis because 0.58 *is above* the threshold of 0.57.

Use two-tailed tests.
Only use one-tailed tests in situations where well-established convention dictates.

We'll see situations where convention calls for one-tailed tests in Part III.

17

Brief on Type I Versus Type II Errors

Your verdict — Unknown truth	Coin is fair (do not reject null hypothesis)	Coin is not fair (reject null hypothesis)
Coin is fair (null hypothesis)	Correct $1 - \alpha$lpha ($\times 100\% =$ confidence)	Incorrect: type I error Concerns truly fair coins αlpha
Coin is not fair	Incorrect: type II error concerns truly unfair coins βeta	Correct $1 - \beta$eta (power)

Note: You may want to reread lesson 1 at this point.

Imagine testing many, many different coins, where their balance is unknown and can be different from coin to coin. Type I error is mistakenly rejecting the null hypothesis when it is actually true. The probability threshold we use for type I error is alpha. When testing a coin for fairness, type I error would be judging a fair coin to be unfair. Type I error is relatively straightforward to control in practice because we set the alpha-level *and we are only concerned with the fate of the fair coins.*

Type II error is mistakenly not rejecting the null hypothesis when it is actually false. With a coin, that would be judging an unfair coin to be fair. The probability for type II error is called beta. We can easily determine beta if we want to test for a certain specific degree of unfairness such as a 30% chance of heads. However, it is difficult to estimate beta in general because then we are *concerned with all the variously unbalanced unfair coins.* You can imagine that for unfair coins that have a 51% chance of coming up heads, then we'll make many type II errors. On the other hand, for unfair coins that have a 1% chance of

Illuminating Statistical Analysis Using Scenarios and Simulations, First Edition.
Jeffrey E Kottemann.
© 2017 John Wiley & Sons, Inc. Published 2017 by John Wiley & Sons, Inc.

coming up heads, then we'll make very few type II errors. What level of unfairness are we dealing with for each coin? We often don't know for sure. We have to make specific assumptions about the level of unfairness. In general, then, we have to make *assumptions* about key population statistic value(s) before we can calculate the probability of type II error.

The complement to beta is statistical power. (Just as alpha and confidence are complements.) Power is the probability of rejecting the null hypothesis when the null hypothesis is in fact false. This is what you want. Power equals 1 minus beta. (Unlike confidence, power is not typically expressed as a percentage, although I will sometimes.)

There are three dynamics we'll consider in depth. (1) As mentioned above, type II error is more likely when an unfair coin is only slightly different than a fair coin. (2) Recall that type I and type II errors are intimately related: If you lower your alpha cutoff in order to lower type I error, then the beta of type II error will go up. (3) The main way to decrease beta and thus increase power for any given alpha-level is to increase the sample size.

1) The simulation results summarized in Table 17.1 show how often type II error occurred—how often unfair coins were erroneously judged fair after being flipped 50 times. Since we are dealing exclusively with unfair coins, the null hypothesis should ideally be rejected all the time. There are separate simulations for various degrees of coin unfairness. Alpha of 0.05 is used: If the proportion of heads in 50 flips falls within the 95% confidence interval for a fair coin, it is judged fair, otherwise it is judged unfair.

Looking at the table, you can see that type II error is indeed common when the unfair coin is close to fair, but becomes less common when the chance of heads gets less and less. Shown in Figure 17.1 is a simulation histogram illustrating the *second* scenario in the table of an unfair coin with 0.40 probability of heads. Visually, the histogram's bell will shift more and more to the left as the chance of heads gets less and less. (Plus, per Chapters 6 and 9, the histogram will condense more and more as the chance of heads gets less and less.)

Table 17.1 Alpha 0.05 50 flips of unfair coins, repeated 10,000 times.

Degree of unfairness: chances of heads	Percentage of type II error
0.45 (Heads 45% of the time on average)	87% (87% of unfair coins judged to be fair)
0.40—see histogram in Figure 17.1	67%—see histogram Figure 17.1
0.30	14%
0.20	0.2%

Statistical power will equal 100% minus the percentages in the right-hand column.

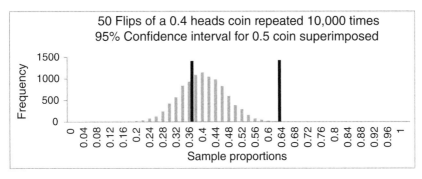

Figure 17.1

2) You can see by Table 17.2 and Figure 17.2 that when we use a 99% confidence interval instead, lowering alpha to 0.01, the confidence interval becomes wider (per Chapter 11) and we'll have even more type II errors. Such widening confidence intervals illustrate that decreasing the alpha-level to increase confidence will cause beta to increase. That is, decreasing the probability of type I error by lowering alpha increases the beta probability of type II error. On the flip side, increasing the alpha-level will decrease the confidence level and narrow the confidence interval; type II error will become less likely.

Table 17.2 Alpha 0.01 50 flips of unfair coins, repeated 10,000 times.

Degree of unfairness: chances of heads	Percentage of type II error
0.45 (Heads 45% of the time on average)	98% (98% of unfair coins judged to be fair)
0.40—see histogram in Figure 17.2	90%—see histogram in Figure 17.2
0.30	43%
0.20	3%

Statistical power will equal 100% minus the percentages in the right-hand column.

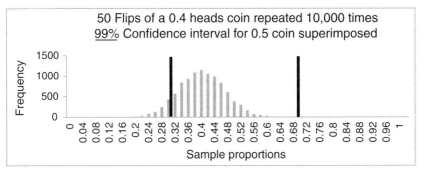

Figure 17.2

> For a given sample size, decreasing the probability of type I error by decreasing the alpha-level (increasing the confidence level) will increase the beta probability of type II error (thus decreasing power). And, vice versa.

3) Next, let's see that increasing the sample size for a given alpha will decrease beta and thus increase power. Comparing the earlier Figure 17.1 histogram with the new Figure 17.3 histogram, you can see that the histogram's bell narrows with the larger sample size, as does the confidence interval (per Chapters 5 and 8). Therefore, as you can see in Table 17.3 compared to Table 17.1, when sample size is increased from 50 to 100, and alpha is kept at 0.05, the percentage of type II error is reduced for each scenario.

Table 17.3 Alpha 0.05 100 flips of unfair coins, repeated 10,000 times.

Degree of unfairness: chances of heads	Percentage of type II error
0.45 (Heads 45% of the time on average)	81% (81% of unfair coins judged to be fair)
0.40—see histogram in Figure 17.3	46%—see histogram in Figure 17.3
0.30	1%
0.20	Nearly zero

Statistical power will equal 100% minus the percentages in the right-hand column.

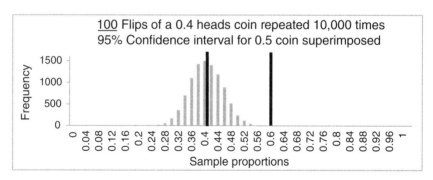

Figure 17.3

Type II errors for the 0.20 *and* 0.30 coins are now both very infrequent. But, with the relatively small disparity between a fair coin and a 0.45 coin as well as between a fair coin and a 0.40 coin, Type II error is still frequent even with a sample size of 100. Evidently we need an even larger sample size to reliably detect these smaller disparities.

> Larger samples can better detect disparities.
> The smaller the disparity, the larger the sample typically needs to
> be to reliably detect it.

There are formulaic methods for these types of analyses that are included in statistical analysis software and online statistics calculators. Researchers use these tools to determine how large their sample sizes need to be to detect what they are looking for.

The Bigger Picture

Imagine that there are 1000 *hypotheses* to be tested and that 100 of the null hypotheses are actually false and 900 of the null hypotheses are actually true. Let's suppose that these are 1000 separate studies to determine whether certain foods and dietary supplements affect peoples' health, and so each null hypothesis states that a certain food or dietary supplement has no effect on health. Assume an alpha-level of 0.05 is used. Also assume that the beta is 0.20 and so statistical power is 0.80. These are fairly realistic levels, although 0.80 power is probably higher than many studies would have.

With a 0.05 alpha probability for type I error, expect $900 \times 0.05 = 45$ type I errors, and expect $900 \times 0.95 = 855$ correct nonrejections of the null hypothesis.

With a 0.20 beta probability for type II error, expect $100 \times 0.20 = 20$ type II errors, and expect $100 \times 0.80 = 80$ correct rejections of the null hypothesis.

Therefore, we expect a total of $45 + 80 = 125$ rejected null hypotheses.

What percentage of the rejected null hypotheses do we expect to be erroneously rejected?

This is sometimes called the <u>false discovery rate</u>. Per the above, we expect 45 to be erroneously rejected, and we expect 80 to be correctly rejected. Therefore, we expect $45/(45 + 80) = 36\%$ of the rejected null hypotheses to be erroneously rejected. Given that rejected null hypotheses tend to get all the publicity, you should find this eye-opening.

We can also lay this out in terms of probabilities as shown in Table 17.4. Using the rightmost column to calculate the proportion of studies rejecting the null hypothesis that are type I errors gives us $0.045 / 0.125 = 0.360 = 36\%$.

The 36% can be improved upon, as you know, by increasing the sample sizes in the 1000 studies. This will increase power. If we increase power to 0.9, for example, we'll increase the correct rejections from 80 to 90. Ideally, if we can increase sample size sufficiently we can set our alpha down to 0.01, thus decreasing the 45 to 9, while also maintaining or even increasing power. With

Table 17.4 Correct and incorrect research study findings.

	Studies not rejecting null	**Studies rejecting null**
Null hypothesis true 90% of the time	$0.9 \times 0.95 = 0.855$ Correct findings	$0.9 \times 0.05 = 0.045$ Type I error
Null hypothesis false 10% of the time	$0.1 \times 0.2 = 0.020$ Type II error	$0.1 \times 0.8 = 0.080$ Correct findings
Column sums	0.875 87.5% of studies will not reject null	0.125 12.5% of studies will reject null

alpha of 0.01, beta of 0.10 and thus power of 0.90, we would reduce the false discovery rate to $9/(9 + 90) = 9\%$. This is much better, but it is also much more expensive to gather the large amounts of additional data that are required.

Lastly, keep in mind that the 36% false discovery rate we came up with is contingent on our assumption that only 100 of the 1000 null hypotheses are actually false. If researchers have good theories to guide their choice of hypotheses, then the proportion of null hypotheses that are truly false should be higher and the false discovery rate should be lower.

This marks the end of Part I. At this point you can look at the review and additional concepts in Part VI Chapter 56, or proceed directly to Part II.

Part II
Sample Means and the Normal Distribution

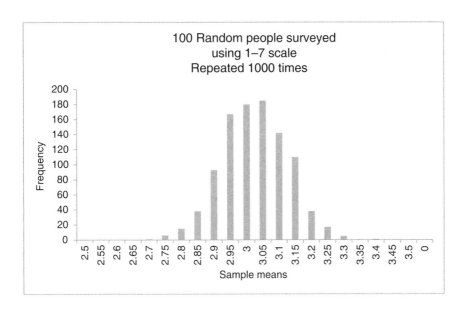

Illuminating Statistical Analysis Using Scenarios and Simulations, First Edition.
Jeffrey E Kottemann.
© 2017 John Wiley & Sons, Inc. Published 2017 by John Wiley & Sons, Inc.

18

Scaled Data and Sample Means

Part II concerns scaled variables—such as heights, weights, incomes, and prices—that take on integer or real number values. Opinions, too, can be expressed on a scale: 1-to-5, or 1-to-7, or 1-to-100, among other options.[1]

The most common summarization of a sample of scaled numbers is the average, what statisticians call the arithmetic mean, or mean for short. The mean is a key sample statistic for scaled variables.

To calculate the mean, you simply total up the sample of numbers and divide by the sample size n.

Below is the sample mean calculation in math shorthand. The symbol for the sample mean is an x with a bar on top, $\bar{x}$ (I will usually spell it out). Σ is the Greek capital letter sigma, which is math shorthand for sum; in this case we are summing a list of n numbers. The sum is then divided by n via $\frac{1}{n}$

$$\bar{x} = \frac{1}{n} \sum_{i=1}^{n} x_i$$

Recall that this same approach to averaging gives us the sample proportions for binomial variables when we represent the individual values with the numbers 0 and 1.

Indeed, *we will find many parallels between analyzing proportions and analyzing means.* The major distinction stems from the following: With binomial variables, proportions are bounded by the values 0 and 1, and proportion variances are bounded by 0 and 0.25. In addition, the proportions themselves dictate proportion variances, with proportion variance equal to $p \times (1 - p)$. With scaled variables, on the other hand, sample means are

---

1 Note to measurement theory experts: In order to continue with the pollsters and public opinion theme, we'll suppose that such response scales give us interval data rather than merely ordinal data. Appendix A overviews types of measurement.

Illuminating Statistical Analysis Using Scenarios and Simulations, First Edition.
Jeffrey E Kottemann.
© 2017 John Wiley & Sons, Inc. Published 2017 by John Wiley & Sons, Inc.

potentially unbounded and sample variances are bounded only on the low end by 0. Both are uncertain estimates; the mean does not dictate the variance.

In Chapter 19, we'll see that sample means are—like sample proportions—normally distributed. Chapters 20 and 21 show again that larger sample sizes and lower variances are associated with lower levels of uncertainty. In Chapters 22–26, we'll see that the determination of the standard error of sample means mirrors that of sample proportions, and the construction of confidence intervals and the derivation of p-values likewise follow from these commonalities. Lastly, in Chapter 27, we'll see that the uncertain nature of sample variances with scaled variables necessitates a tweak to the normal distribution for purposes of statistical inference: the z-distribution is replaced with the t-distribution.

19

Distribution of Random Sample Means

<table>
<tr><td>Statistical Scenario—Sample Means with Random Numbers</td></tr>
<tr><td>

Fifty random monkeys, each randomly pick a number from the following list of values.

1   2   3   4   5   6   7

Then you add up the 50 numbers and calculate the sample mean (sum/n).

What is the best guess for what the sample mean will be? (Answer: 4)

Say you do this 1000 times and put the 1000 sample means in a histogram. What will the histogram look like?

</td></tr>
</table>

The histogram in Figure 19.1 summarizes the corresponding simulation results.

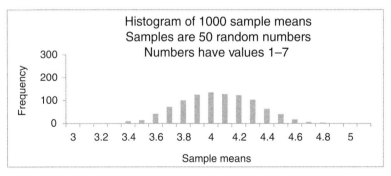

Figure 19.1 A histogram bar is the count of sample means that are less than or equal to the bar's number label, and strictly greater than the number label of the bar to its left. So, for example, the 3.6 bar is the number of sample means that are ≤3.6 and >3.5.

Illuminating Statistical Analysis Using Scenarios and Simulations, First Edition.
Jeffrey E Kottemann.
© 2017 John Wiley & Sons, Inc. Published 2017 by John Wiley & Sons, Inc.

The population mean for random numbers 1–7 is 4, and the distribution of sample means is normal and centered on the population mean.[1]

Just as it is unlikely that a fair coin will come up tails nearly all the time or heads nearly all the time, it is unlikely that the great majority of 50 random numbers will be less than 4 and it is unlikely that the great majority will be greater than 4. In general, it is unlikely to get a random sample with the great majority of its values below the population mean and it is unlikely to get a random sample with a great majority of its values above the population mean.

The expected value of a sample mean is the population mean.

Sample means will be normally distributed around the population mean.[2]

---

1 The Greek letter mu is used to designate the *population* mean, μ, but I will continue to spell it out.

2 The central limit theorem states that distributions of sample means become normal with adequate sample size. A common rule of thumb is that sample size should be at least 30. In practice, sample size adequacy depends on the actual distribution of the data itself as well as the type of statistical analysis method being used. We'll wait until Part V to consider this in more detail.

20

Amount of Evidence

The original histogram, reproduced in Figure 20.1, shows the distribution of random sample means for a sample size of 50.

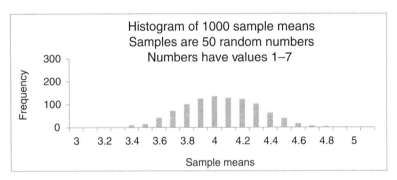

Figure 20.1

Figure 20.2 is a histogram showing the distribution of sample means when sample size is 200. Like before, we use random integers 1–7. Notice that it is more compact.

Illuminating Statistical Analysis Using Scenarios and Simulations, First Edition.
Jeffrey E Kottemann.
© 2017 John Wiley & Sons, Inc. Published 2017 by John Wiley & Sons, Inc.

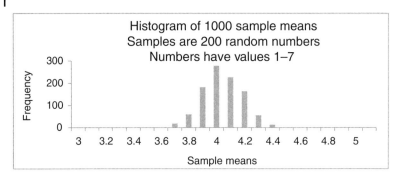

Figure 20.2

Analogous to what we saw with sample proportions:

When sample sizes are larger,
sample means tend to be closer to the population mean.[1]
More evidence (larger sample sizes) → less uncertainty

As we'll see, the uncertainty due to sample size n is incorporated into various statistical formulas involving sample means.

---

1 As noted in lesson 5, this principle is called the Law of Large Numbers.

21

Variance of Evidence

Statistical Scenario—Variance of Evidence

When variance decreases, will the sample mean distribution get more compact like it did with sample proportions?

The original histogram, reproduced in Figure 21.1, shows the distribution of random sample means for a sample size of 50 when data values are 1–7.

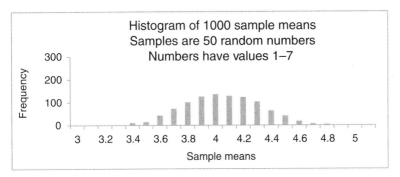

Figure 21.1

Figure 21.2 is a histogram of sample means with random samples of the same size but the data only take on the values 3–5. The resulting distribution is more compact.

Illuminating Statistical Analysis Using Scenarios and Simulations, First Edition.
Jeffrey E Kottemann.

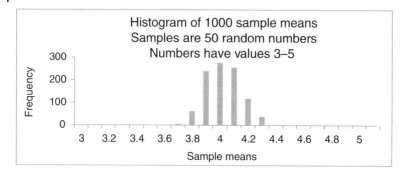

Figure 21.2

Analogous to what we saw with sample proportions:

> When variance is lower,
> sample means tend to be closer to the population mean.

> Lower variance → less uncertainty

Recall that a key statistic to summarize variety is appropriately named the variance. And for a sample it is sample variance. Expressed in words, the sample variance of scaled data is the average squared distance of the sample values from the sample mean. It captures how spread out a sample's data values are.

To get the sample variance, we do the following: For each sample value, subtract the sample mean from it and square the difference; add the result to a running total. Note that we have to square the distances to make them all positive, otherwise the positive and negative distances would cancel each other out. After we are done with that, we divide the total by the sample size minus 1. (Don't worry about why 1 is subtracted from the sample size for now.)

Below is the sample variance calculation in math shorthand.[1] The symbol for sample variance is s^2 (I will usually spell it out).

$$s^2 = \frac{1}{n-1} \sum_{i=1}^{n} (x_i - \bar{x})^2$$

---

1 As footnoted in Part I Chapter 6, when x is a binomial variable with values 0 or 1, we can derive the specialized formula for the proportion variance using the general formula for population variance.

$$\frac{1}{n} \sum_{i=1}^{n} (x_i - \bar{x})^2 = p \times (1-p)$$

Let's think once again about how variance is related to uncertainty. Consider two community populations of people and their answers to a 1-to-7-scaled strongly disagree to strongly agree opinion survey question. In the first community, imagine that all the people are indifferent; everyone's opinion is 4. Population variance is zero. In the second community, people have a variety of opinions 1, 2, 3, 4, 5, 6, or 7. Population variance is greater than zero. Let's say that the *population mean* for both communities is 4.

Now we take a random sample of 100 from each community. The sample mean for the first community will be 4 and we'll be quite certain that 4 is indeed the population mean because our sample variance will be zero. For the second community, the sample mean will most likely be close to 4 but not exactly equal to 4, and we'll be less certain that our sample mean is correct because our sample variance will be greater than zero (barring the *extremely* unlikely event of having all 100 opinions in the random sample from the second community be the same).

Ponder the thesaurus entries for these three highly related common words.

Variety: diversity, variability, variation.
Variance: discrepancy, inconsistency, variation.
Inconsistency: discrepancy, unpredictability, variation.

Sample data is evidence that we use in drawing conclusions. Sample data with higher variance gives us less consistent (more varied) evidence and therefore our conclusions based on the evidence are less certain. On the other hand, sample data with lower variance gives us more consistent (less varied) evidence and therefore our conclusions based on the evidence are more certain.

As we'll see, the uncertainty due to sample variance s^2 is incorporated into various statistical formulas involving sample means.

Both sample size and sample variance are sources of uncertainty.

Variance and Standard Deviation

Since sample variance s^2 involves the *squared* distances of the sample values from the sample mean, its unit is also the square. For example with a 1-to-7 opinion question, the variance unit is opinion squared—an unintuitive unit. For this and other reasons, the square root of variance is sometimes used. It is called the <u>standard deviation</u> and is symbolized by s. I will occasionally refer to standard deviation, but we'll explicitly use variance almost exclusively. Moreover, in Part II Chapter 27 & Part III we'll take a close look at the statistical behavior of sample variances in their own right.

22

Homing in on the Population Mean I

Let's look at some simulation results that mimic 1000 surveyors who each survey 100 randomly selected people in the community—Figure 22.1. Only omniscient beings know that the overall community average is 4.0 and the overall community variance is 1.0.

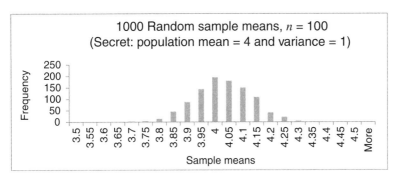

Figure 22.1

Illuminating Statistical Analysis Using Scenarios and Simulations, First Edition.
Jeffrey E Kottemann.
© 2017 John Wiley & Sons, Inc. Published 2017 by John Wiley & Sons, Inc.

> What are the chances that a surveyor would get a sample mean of 4.5?

It looks like it is almost impossible for a surveyor to get a sample that has a sample mean of 4.5. In other words, with a sample size of 100 it seems that sample means will almost never be that far from the true population mean; random samples will almost never be that out of whack. About 95% of the simulation's sample means are within 0.2 of the true population mean.

Next, let's look at the formulaic method. There is a "trick" formula that has the same form as the one we used for proportions. Standard Error in both cases is the square root of the variance divided by the sample size.

Recall the 95% confidence interval formula we used for sample proportions in Chapter 7.

$$\text{From } p - 1.96\sqrt{(p \times (1 - p))/n} \text{ to } p + 1.96\sqrt{(p \times (1 - p))/n}$$

$\sqrt{(p \times (1 - p))/n}$ is the standard error of a sample proportion involving the proportion variance $(p \times (1 - p))$ and the sample size n.

Analogously, below is the 95% confidence interval formula for sample means.

> $$\text{From mean} - 1.96\sqrt{(\sigma^2/n)} \text{ to mean} + 1.96\sqrt{(\sigma^2/n)}$$

$\sqrt{(\sigma^2/n)}$ is the <u>standard error of a sample mean</u>[1] involving the population variance, symbolized by σ^2, and the sample size n.

Plugging in the *statistical scenario* numbers we get the 95% confidence interval for the population mean:

$$\text{From } 4.0 - 1.96\sqrt{1/100} \text{ to } 4.0 + 1.96\sqrt{1/100}$$

$$\text{From } 4.0 - 0.196 \text{ to } 4.0 + 0.196$$

That is from 3.804 to 4.196, which does not contain 4.5. So, it seems very unlikely that a surveyor would get a sample that has a sample mean of 4.5, which is the same conclusion we reached with the simulation results. The 95%

---

1 The Greek lowercase letter sigma is used to designate the *population* variance, σ^2, and the *population* standard deviation, σ. I will only use the symbol occasionally. Usually, I will continue to spell things out.

confidence interval derived via the formula is about 0.2 on each side, which agrees with the 0.2 found with the simulation.[2]

Next, we'll adopt a single surveyor's perspective, and we will use a sample mean that is closer to the population mean.

---

2 At some point you should look at Appendix C: standard error as standard deviation.

23

Homing in on the Population Mean II

Let's look at this from the perspective of one of the surveyors (you).

Statistical Scenario—Sample Mean II

Say you want to see how strongly, on average, members of your community agree or disagree with a new proposed policy. You decide to survey 100 random members and record their responses using the following scale:

Strongly agree   1   2   3   4   5   6   7   Strongly disagree

Say your sample mean is 4.15 and your sample variance is 1.0.

Could the true community mean equal 4.0?

In other words, does the 95% confidence interval centered on the sample mean of 4.15 contain the value 4.0?

As the lone surveyor, you will want to construct a 95% confidence interval around your sample of 4.15 to check whether it contains 4.0. Since the true population variance in this scenario is unknown—as is usually the case in practice—the sample variance is used instead.

Note: Using the sample variance as an estimate of the population variance introduces additional uncertainty. We won't address this source of uncertainty until Chapter 27.

Let's simulate the uncertainty surrounding our sample mean of 4.15. We'll simulate 1000 random samples of size 100 based on the scenario's mean of 4.15 and variance of 1.0. Given the results shown in Figure 23.1, it appears that 4.0 is not especially unlikely. Eyeballing, the 95% confidence interval appears to be about 3.95–4.35. Given this, we would not rule out the hypothesis that the true population mean could be 4.0.

Illuminating Statistical Analysis Using Scenarios and Simulations, First Edition.
Jeffrey E Kottemann.
© 2017 John Wiley & Sons, Inc. Published 2017 by John Wiley & Sons, Inc.

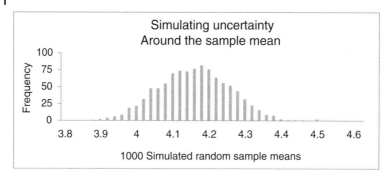

Figure 23.1

(Alternatively, we could have simulated the case of a population mean of 4.0 with a variance of 1.0 and seen that 4.15 is not especially unlikely. In effect, this just shifts the histogram over to be centered on 4.0 instead.)

Now let's use the formulaic approach.

$$\text{From sample mean} - 1.96\sqrt{(s^2/n)} \text{ to sample mean} + 1.96\sqrt{(s^2/n)}$$

$\sqrt{(s^2/n)}$ is the estimate for the standard error of a sample mean involving the sample variance s^2 and the sample size n.

It is only an *estimate* of standard error because we are using the sample variance rather that the population variance, and the sample variance is an estimate of the population variance. It is usually simply referred to as standard error though.

Plugging in the *statistical scenario* numbers we get a 95% confidence interval:

$$\text{From } 4.15 - 1.96\sqrt{1/100} \text{ to } 4.15 + 1.96\sqrt{1/100}$$

That is from 3.954 to 4.346, which does contain 4.0. This agrees with the simulation results.

Do not reject the hypothesis that the true mean could equal 4.0.

(Alternatively, we could have used a mean of 4.0 with a variance of 1.0 to get a confidence interval of 3.804–4.196, which contains 4.15.)

24

Homing in on the Population Mean III

Statistical Scenario—p-Values

Again, say your sample mean is 4.15 and your sample variance is 1.0.

Could the true community mean be 4.0 or less?

What is the *p*-value?

In the previous lesson, we used a confidence interval to justify not rejecting the null hypothesis. If we want to be more specific than "we cannot rule it out at the 95% confidence level" we can calculate a *p*-value.

Just as constructing confidence intervals for means is analogous to what we did for proportions, deriving *p*-values for means is analogous to what we did for proportions.

Eyeballing our previous histogram, reproduced in Figure 24.1, we can add up the frequencies of all the sample means less than or equal to 4. It looks to be about 70 total. Two-tailed, it is 140. Dividing by 1000 to get relative frequency, we get 0.14. That is our rough approximation for the two-tail *p*-value, which reflects the probability of the population mean being 0.15 or more away from 4.15 on either side.

Illuminating Statistical Analysis Using Scenarios and Simulations, First Edition.
Jeffrey E Kottemann.
© 2017 John Wiley & Sons, Inc. Published 2017 by John Wiley & Sons, Inc.

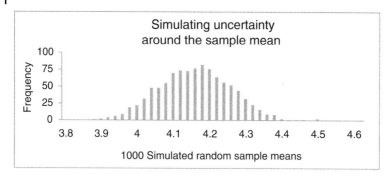

Figure 24.1

Now let's use the formulaic approach.

First we need to determine the number of standard errors that fit into the actual difference of $4.0 - 4.15 = -0.15$. Standard error is $\sqrt{(s^2/n)}$ that gives us $\sqrt{(1/100)} = 0.1$. Then, -0.15 divided by 0.1 equals -1.5 standard errors. The p-value for -1.5 standard errors calculated with a statistical calculator is 0.134 for two-tails. That is the area of the standard normal z-distribution from negative infinity to -1.5 plus the area from $+1.5$ to positive infinity. Figure 24.2 illustrates.

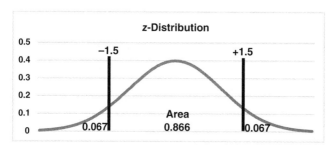

Figure 24.2

Could the true community average equal 4.0 or less (or 4.3 or more)?

We cannot rule it out. The p-value equals 0.134, two-tail. This is not strong enough evidence to reject the null hypothesis when using an alpha-level of 0.05, or even the more relaxed alpha-level of 0.10. This, as usual, agrees with the simulation results.

25

Judging Mean Differences

Statistical Scenario—Sample Mean Difference

Say you want to see if there is a difference of opinion between females and males in your community regarding a new proposed policy. You decide to survey 100 random women and 100 random men and record their responses using the following scale:

Strongly agree   1   2   3   4   5   6   7   Strongly disagree

This is your evidence:
Your sample mean for females is 3.0 with a sample variance of 3.1.
And your sample mean for males is 3.5 with a sample variance of 2.9.

How likely is it to get a sample mean difference as extreme as −0.5 (and, two-tail, also +0.5) by chance when males and females have the same population mean?

Null hypothesis: The female population mean minus the male population mean equals zero. In other words, the female and male population means are equal.

This *statistical scenario* is analogous to what we saw with the difference between two sample proportions. For the simulation, we'll assume the null hypothesis of equal population means for females and males is true, so the difference in the population means is zero. We'll make both the males' and females' population means equal 3.0. We want to see how frequently we should expect to get two sample means that are 0.5 apart or more simply by chance when the two population means are in fact equal. The uncertainty, as usual, is driven by the variances and sample sizes. We'll use the sample variances of 3.1 and 2.9 as estimates for the female and male population variances. The sample sizes are 100 and 100.

Illuminating Statistical Analysis Using Scenarios and Simulations, First Edition.
Jeffrey E Kottemann.
© 2017 John Wiley & Sons, Inc. Published 2017 by John Wiley & Sons, Inc.

We'll simulate 10,000 pollsters to help ensure that we get a dense, smooth histogram. When looking at the histogram, we want to see how many of the 10,000 sample mean differences are less than or equal to −0.5. We'll simply double that to get two-tail. This will tell us how likely it is to get a female sample mean and a male sample mean that are 0.5 apart from each other (or more) when in fact there is no difference between the female and male population means. Figure 25.1 shows the simulation results.

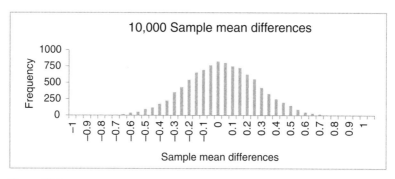

Figure 25.1

Eyeballing the histogram bars in Figure 25.1, a total of approximately 200 of the 10,000 sample mean differences are less than or equal to −0.5, equating to an approximate one-tail *p*-value of 0.02. The two-tail *p*-value estimate is double that, equaling 0.04. Using an alpha-level of 0.05, we will reject the null hypothesis of no population mean difference. As we'll see next, this is what the formulaic approach gives us too.

Basically, we need to see how many standard errors fit into the actual sample mean difference of 3.0-3.5 = −0.5 (or 3.5-3.0 = +0.5). Expressed as one formula with custom labeling for female and male we have

$$\frac{\text{sample mean}_F - \text{sample mean}_M}{\sqrt{\dfrac{s_F^2}{n_F} + \dfrac{s_M^2}{n_M}}} = \frac{\text{sample mean difference}}{\text{standard error}} = \#SEs$$

The new aspect here is how the standard error is calculated when we have two samples. Earlier, we saw that the standard error for one sample mean is $\sqrt{(s^2/n)}$. Now we have two samples, each with its own variance and size, which are incorporated into the formula for the <u>standard error of the difference between two sample means</u>. Notice that the formula is similar in structure to the formula for proportion differences (Chapter 13).

Plugging in the numbers to calculate standard error gives $\sqrt{(3.1/100 + 2.9/100)} = 0.245$ rounded.

The number of standard errors of this size (0.245) that fit into the actual sample mean difference (-0.5) gives us $-0.5/0.245 = -2.04$ rounded. This rescales the actual difference into standard error difference. (Or we could say we standardized the difference.)

Plugging the #SEs value of -2.04 into a statistical calculator gives a p-value of 0.04 (rounded) for two-tails ($0.02 + 0.02$). That is the z-distribution area from negative infinity to -2.04 plus the area from $+2.04$ to positive infinity. Figure 25.2 illustrates. As expected, this p-value agrees with the simulation results.

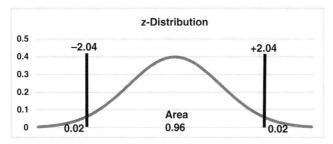

Figure 25.2

If we are using a 0.05 alpha-level, then we would say the sample mean difference between females and males is statistically significant because the p-value is less than or equal to our 0.05 alpha-level. We would reject the null hypothesis of equal female and male population means. (If we were using the stricter 0.01 alpha-level, then we would say the sample mean difference is not statistically significant and we would not reject the null hypothesis.)

26

Sample Size, Variance, and Uncertainty

Statistical Scenarios—Assorted Sample Sizes and Variances

Let's review:
With our investigation, what happens as we increase or decrease our sample sizes?
What happens if the sample variances happen to be higher or lower?

Let's see how various changes to the "base case" of the previous chapter impact the original p-value of 0.04. We'll change one aspect at a time, leaving all the others as they were. The formulaic calculations and the simulation histogram are given for each case. For all the simulations, we'll assume the null hypothesis of equal population means for females and males is true, so the difference in the population means is zero. We'll make both the males' and females' population means equal 3.0. We want to see how frequently we should expect to get two sample means that are 0.5 apart or more simply by chance when the two population means are in fact equal. The uncertainty, as usual, is driven by the variances and sample sizes. We'll use an alpha-level of 0.05 (95% confidence level) for declaring statistical significance. Let's start by reviewing the base case.

(A) Base case.

The sample variances are 3.1 and 2.9, and the sample sizes are 100 and 100.

$\sqrt{(3.1/100) + (2.9/100)} = 0.245$ for standard error

$-0.5/2.45 = -2.04$ standard errors fit into the actual difference

p-value $= 0.04$ (rounded)

Reject the null hypothesis of no difference between population means

Illuminating Statistical Analysis Using Scenarios and Simulations, First Edition.
Jeffrey E Kottemann.
© 2017 John Wiley & Sons, Inc. Published 2017 by John Wiley & Sons, Inc.

Figure 26.1 shows simulation results. Sample mean differences outside the −0.5 to +0.5 interval are relatively rare when the null hypothesis of equal population means is true.

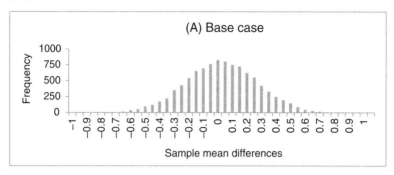

Figure 26.1

First, let's look at different sample sizes.

(B) More evidence, less uncertainty.

Let's recalculate using bigger sample sizes of 500 random women and 500 random men instead of 100 and 100.

$\sqrt{(3.1/500) + (2.9/500)} = 0.110$ for standard error

$-.05/0.110 = -4.564$ standard errors fit into the actual difference
p-value $= 0.00001$ (rounded)
Reject the null hypothesis of no difference between population means

Figure 26.2 shows simulation results using the larger sample sizes. As expected, the distribution of sample mean differences has become less spread out compared to the base case. Because of this, sample mean differences outside the −0.5 to +0.5 interval are now extremely rare when the null hypothesis of equal population means is true.

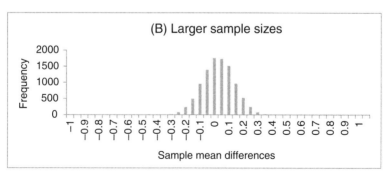

Figure 26.2

(C) Less evidence, more uncertainty.

Let's recalculate using smaller sample sizes of 50 and 50 instead of 100 and 100.

$\sqrt{(3.1/50) + (2.9/50)} = 0.346$ for standard error

$-0.5/0.346 = -1.443$ standard errors fit into the actual difference
p-value $= 0.15$ (rounded)
Do not reject the null hypothesis of no difference between population means

Figure 26.3 shows simulation results using the smaller sample sizes. As expected, the distribution of sample mean differences has become more spread out compared to the base case. Because of this, sample mean differences outside the −0.5 to +0.5 interval are now fairly common when the null hypothesis of equal population means is true.

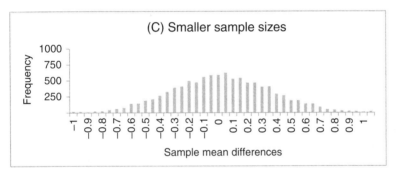

Figure 26.3

Next, let's look at different sample variances.

(D) More variance, more uncertainty.

Let's recalculate using larger sample variances of 8.0 and 8.0 instead of 3.1 and 2.9. We'll use the original sample sizes of 100 and 100.

$\sqrt{(8.0/100) + (8.0/100)} = 0.400$ for standard error

$-0.5/0.400 = -1.250$ standard errors fit into the actual difference
p-value $= 0.21$ (rounded)
Do not reject the null hypothesis of no difference between population means

Figure 26.4 shows simulation results using the larger variances. As expected, the distribution of sample mean differences has become more spread out compared to the base case. Because of this, sample mean differences outside the −0.5 to +0.5 interval are now fairly common when the null hypothesis of equal population means is true.

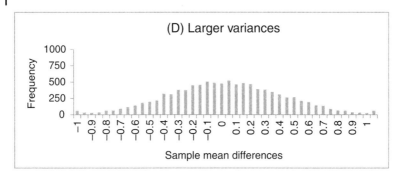

Figure 26.4

(E) Less variance, less uncertainty.

Let's recalculate using smaller variances of 1.0 and 1.0 instead of 3.1 and 2.9. We'll use the original sample sizes of 100 and 100.

$\sqrt{(1.0/100) + (1.0/100)} = 0.141$ for standard error

$-0.5/0.141 = -3.536$ standard errors fit into the actual difference

p-value $= 0.0004$ (rounded)

Reject the null hypothesis of no difference between population means

Figure 26.5 shows simulation results using the smaller variances. As expected, the distribution of sample mean differences has become less spread out compared to the base case. Because of this, sample mean differences outside the −0.5 to +0.5 interval are now very rare when the null hypothesis of equal population means is true.

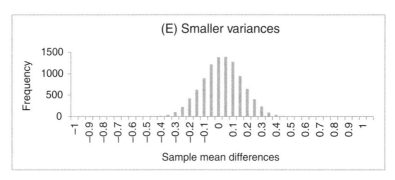

Figure 26.5

Table 26.1 and the two-tail z-distribution diagram in Figure 26.6 summarize the five different cases.

Table 26.1 Summary of the five cases.

		s_F^2	s_M^2	n_F	n_M	Mean difference	Standard error (SE)	#SEs (z)	Two-tail p-value
A	Base case	3.1	2.9	100	100	−0.5	0.24495	−2.041	0.04123
B	Larger n	3.1	2.9	500	500	−0.5	0.10954	−4.564	0.00001
C	Smaller n	3.1	2.9	50	50	−0.5	0.34641	−1.443	0.14891
D	Larger s^2	8.0	8.0	100	100	−0.5	0.40000	−1.250	0.21130
E	Smaller s^2	1.0	1.0	100	100	−0.5	0.14142	−3.536	0.00041

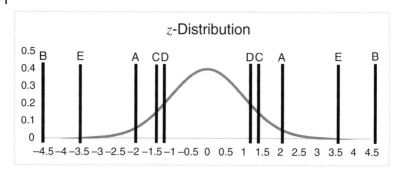

Figure 26.6

To look at these dynamics, I could have saved myself a lot of tedious work by just logically interpreting the formula.

$$\frac{\text{sample mean}_F - \text{sample mean}_M}{\sqrt{\dfrac{s_F^2}{n_F} + \dfrac{s_M^2}{n_M}}} = \frac{\text{sample mean difference}}{\text{standard error}} = \#\text{SEs}$$

As the actual difference between the sample means increases, #SEs will move further from zero, and the p-value will get smaller. And, vice versa. The sign of #SEs depends solely on which of the two sample means is biggest, and only matters for one-tailed assessments (such as hypothesizing that the female population mean is less than the male population mean).

As the sample size gets larger, the standard error gets smaller, so #SEs will move further from zero, and the p-value will get smaller. More evidence implies less uncertainty. And, vice versa.

As the variance gets smaller, the standard error gets smaller, so #SEs will move further from zero, and the p-value will get smaller. Less variance implies less uncertainty. And vice versa.[1]

Possible reasons for nonsignificant results (p-values > alpha):

The null hypothesis is actually true (the population means are in fact equal).
The null hypothesis is actually false, but sample size is too small.
The null hypothesis is actually false, but sample variance happens to be large.

---

1 In the *extremely* unlikely case that both sample variances are zero, then standard error will be zero. In such a case, there is no uncertainty attached to the sample means, and so there is no reason to use statistical analysis. (Besides, dividing by zero is mathematically undefined.) However, if in fact you do get sample variances equal to zero, you must wonder if you are doing something wrong.

27

The *t*-Distribution

In Chapter 22, we used the population variance to calculate the standard error of a sample mean via $\sqrt{\sigma^2/n}$. In practice, we rarely know the population variance and so in subsequent chapters we used the sample variance instead to calculate an estimate of the standard error of a sample mean via $\sqrt{s^2/n}$. But, as noted in Chapter 23, this introduces additional uncertainty, particularly when we have small sample sizes.

Let's look at the results of four random sampling simulations involving sample sizes of 4, 7, 60, and 1000. Each of the four simulations randomly samples from a population with population mean of 10 and population variance of 4 (there is nothing special about these values). Results are shown in Figures 27.1–27.4. The figure "a" histograms show the distributions of the sample variances, s^2, for each of the four sample sizes. The corresponding figure "b" histograms show the distributions of the #SEs calculated via $(\bar{x} - 10)/\sqrt{s^2/n}$, which is the difference between a sample mean and the population mean rescaled by the standard error estimate. The horizontal axes end at −1.96 and +1.96 to make it easier to count how many #SEs are out in the tails. The histogram bar at #SEs of −1.96 is for all #SEs less than or equal to −1.96, and the bar to the right of 1.96 is for all #SEs greater than +1.96.

Notice two things:

1) *The figure "a" sample variance histograms*: When sample sizes are smaller, sample variances are more uncertain estimates of the population variance. For example, with $n = 4$, the sample variances range widely (and nonnormally) around the population variance of 4. In contrast, with $n = 1000$, all the sample variances are very close to 4 (and the distribution is almost perfectly normal; that cannot be seen because of the scale of the horizontal axis).[1]

2) *The figure "b" sample #SEs histograms*: When we use sample variances instead of population variances to calculate #SEs, the #SEs distributions

---

1 Sample variances form X^2 distributions, which we will see in Part III.

Illuminating Statistical Analysis Using Scenarios and Simulations, First Edition.
Jeffrey E Kottemann.
© 2017 John Wiley & Sons, Inc. Published 2017 by John Wiley & Sons, Inc.

(a)

(b)

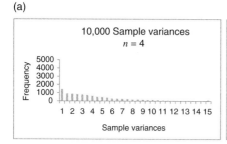

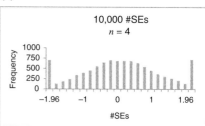

Figure 27.1

(a)

(b)

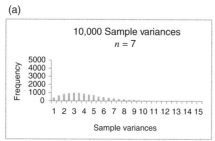

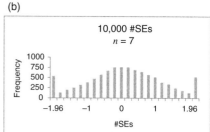

Figure 27.2

(a)

(b)

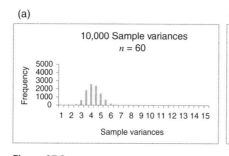

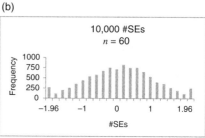

Figure 27.3

(a)

(b)

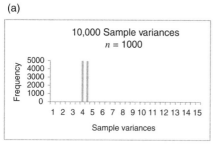

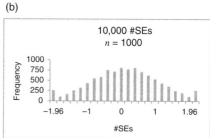

Figure 27.4

widen and deform from the normal shape, especially with smaller sample sizes. For example, with $n = 4$, nearly 15% of the #SEs are outside the interval -1.96 to $+1.96$. That is certainly not indicative of a normal distribution. In contrast, with $n = 1000$, the distribution of #SEs has become almost perfectly normal with 5% of the #SEs outside the interval -1.96 to $+1.96$.

It is evident from the simulation results that sample variances are uncertain estimates of population variances, especially with smaller samples. And, when sample variances are used to determine standard error, the #SEs distributions widen and deform from the normal shape, especially with smaller samples. Because of this, an alternative to the normal z-distribution had to be developed: the t-distributions. Mirroring the #SEs simulation results, t-distributions change shape as a function of sample size.

Figure 27.5 is a graphic showing four different t-distributions superimposed together. They are continuous distributions corresponding to the four #SEs histograms. They are somewhat stylized to accentuate the differences between them.[2]

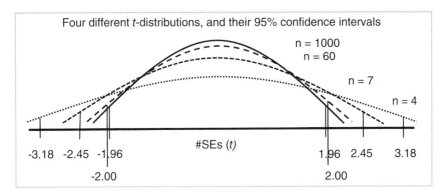

Figure 27.5

The t-distributions systematically adjust as a function of sample size. The adjustments yield wider confidence intervals and larger p-values than does the z-distribution, especially for smaller samples. Looking at the figure, notice that with a sample size of 4, the 95% confidence interval is 3.18 standard errors rather than the z-distribution's 1.96 standard errors. As sample size increases, t-distributions becomes more and more normal. When sample size is 60, the t-distribution is nearly normal (± 2.00 SE versus ± 1.96 SE). When sample size is 1000, the difference is nearly nonexistent.

Table 27.1 adds t-distribution p-values to the table from the previous chapter. You can compare the p-values we get with the z-distribution to the p-values we

---

2 You can see the formula here: http://itl.nist.gov/div898/handbook/eda/section3/eda3664.htm

Table 27.1 Summary of the five cases using z- and t- distributions.

		s_F^2	s_M^2	n_F	n_M	Mean difference	SE	#SEs	z-Distribution p-value	t-Distribution p-value
A	Base case	3.1	2.9	100	100	−0.5	0.24495	−2.041	0.04123	0.04255
B	Larger n	3.1	2.9	500	500	−0.5	0.10954	−4.564	0.00001	0.00001
C	Smaller n	3.1	2.9	50	50	−0.5	0.34641	−1.443	0.14891	0.15210
D	Larger s^2	8.0	8.0	100	100	−0.5	0.40000	−1.250	0.21130	0.21277
E	Smaller s^2	1.0	1.0	100	100	−0.5	0.14142	−3.536	0.00041	0.00051

get with t-distributions. The t-distributions' p-values are slightly larger, and more so with smaller samples. With the bigger samples of case B there is no noticeable difference. P-values with the z-distribution are determined simply by #SEs; p-values for t-distributions are determined by #SEs along with sample sizes.[3]

When we use #SEs to get p-values using t-distributions, #SEs also go by the designation t. In other words, "#SEs" and "t" are synonymous when we are using #SEs to reference t-distributions. It is a labeling convention. Recall that the same general convention was used with #SEs, z, and the z-distribution (Chapter 15). And, just as there are p-value calculators for z values, there are p-value calculators for t values.

> With scaled data, using the sample variance to calculate standard error introduces additional uncertainty into statistical inference. The smaller the sample size, the more the uncertainty. The t-distributions account for this source of uncertainty and should be used in place of the z-distribution.

Everything we have done in Part II still applies, it is just that instead of using the z-distribution to determine confidence intervals and performing z-tests to get p-values, we will use t-distributions to determine confidence intervals and perform analogous t-tests to get p-values. We'll be seeing t-distributions again, but not until Part IV. Don't forget about them.

> This marks the end of Part II. At this point you can look at the review and additional concepts in Part VI Chapter 57, or proceed directly to Part III.

---

3 It is specifically determined by "degrees of freedom" rather than simply sample size. The principles of degrees of freedom are covered in Chapter 29.

Part III
Multiple Proportions and Means: The X^2- and F-Distributions

Chapters 28–32 expand from comparing the sample proportions of two groups to comparing the sample proportions of multiple groups.

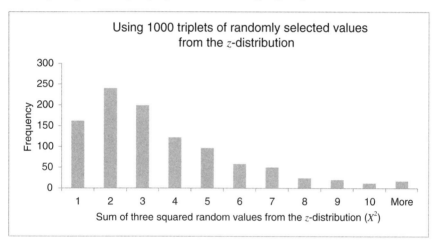

Illuminating Statistical Analysis Using Scenarios and Simulations, First Edition.
Jeffrey E Kottemann.
© 2017 John Wiley & Sons, Inc. Published 2017 by John Wiley & Sons, Inc.

Chapter 33 uses variance ratios to compare variances for scaled data.
Chapters 34 and 35 use variance ratios to expand from comparing the sample means of two groups to comparing the sample means of multiple groups.

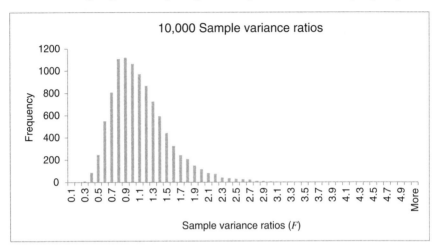

28

Multiple Proportions and the X^2-Distribution

Statistical Scenario—Political Affiliation

You are wondering about political affiliations in your community, so you randomly survey 90 community members and ask each one if they are a Democrat, a Republican, or an Independent.

Your null hypothesis is that your community's political affiliations are perfectly balanced (evenly split), with one-third in each category.

Under the null hypothesis, your expected sample counts are 30, 30, and 30.

Say that the actual sample counts you observe turn out to be 40, 30, and 20.

Should you reject your null hypothesis?

How likely is it to randomly sample 40, 30, 20 from an evenly split population?

Let's expand from scenarios involving binomial variables, as we investigated in Part I, to scenarios involving multinomial variables. Political affiliation is a multinomial variable with three possible values: D, R, and I. How would we go about assessing whether a community is evenly split between the three, with 33.3% in each category? You have a random sample of 90 community members with 40 D, 30 R, and 20 I. If the null hypothesis is true, we expect about 30, 30, and 30 sample members in each of the three political affiliation categories. The word "about" needs to be more exact: would 35, 30, 25 be close enough? 37, 30, 23? 40, 30, 20?

We need to (re)invent a sample statistic that reflects how far the actual observed counts are from our expected counts. If we sum up the differences between the observed sample counts and the expected counts, that might tell us how far apart the observed and expected values are $(40\text{-}30) + (30\text{-}30) + (20\text{-}30) = 0$. Whoops! if we simply add up the differences, the positive and negative differences will cancel each other. We'll have to do what is usually done, we'll square the differences to make them all positive (reminiscent of

Illuminating Statistical Analysis Using Scenarios and Simulations, First Edition.
Jeffrey E Kottemann.
© 2017 John Wiley & Sons, Inc. Published 2017 by John Wiley & Sons, Inc.

what we did in calculating sample variance). This gives us $(40\text{-}30)^2 + (30\text{-}30)^2 + (20\text{-}30)^2$. Now, we'll want these squared count differences rescaled to proportions: $(40\text{-}30)^2/30 + (30\text{-}30)^2/30 + (20\text{-}30)^2/30$. This gives us the squared count differences as proportions of our expected counts.

$$\sum \frac{(\text{observed} - \text{expected})^2}{\text{expected}} \text{ of all the categories} = \text{the } \chi^2 \text{ statistic}$$

We now have a useable sample statistic. I'll denote it X^2 for convenience. Officially, the Greek lower case letter is used: χ^2 is called Chi-squared; Chi is pronounced like "kite." I'll use the symbol X^2 rather than χ^2 in the text because χ^2 hangs down and messes up line spacing.

Our scenario has $X^2 = 3.33 + 0 + 3.33 = 6.67$. How likely is this X^2 value to arise by chance? We need a corresponding distribution for chance occurrences of X^2. The distributions we'll reinvent will be X^2-distributions, called Chi-squared distributions.

As we saw in Part I, sample proportions are normally distributed as are sample proportion differences. X^2 involves proportions with squared terms, so first let's see what we get when we take a single *random* value from the standard normal distribution (z-distribution) and square it: giving us a random z^2. We'll do that 1000 times and make a histogram. Second, let's see what we get when we take the sum of two randomly selected values from the z-distribution that have each been squared. We'll do that 1000 times. Third, let's see what we get when we take the sum of three randomly selected values from the z-distribution that have each been squared. We'll do that 1000 times. We could go on, but that is all we need for this chapter's scenario.

Table 28.1 shows examples of the numbers we are using. Since the values are randomly selected according to the z-distribution, 95% of the randomly selected values will be between -1.96 and $+1.96$, and 99% will be between -2.58 and $+2.58$.[1]

Table 28.1 Some random z-values and their squared values.

Some random numbers from the z-distribution				
0.128402	1.131352	−0.960734	0.927008	−0.339121
The random numbers from the z-distribution squared				
0.016487	1.279957	0.923010	0.859345	0.115003

---

1 A random value, z, from the standard normal distribution is sometimes called a standard normal deviate, meaning a deviation from zero using the standard normal distribution (the z-distribution). The relationship between a single z^2 and a single X^2 term derives from $z = \frac{\text{observed} - \text{expected}}{\sqrt{\text{expected}}}$ where squaring both sides gives $z^2 = \frac{(\text{observed} - \text{expected})^2}{\text{expected}}$.

Figure 28.1 is a histogram of 1000 simulation results when we randomly select just one value from the z-distribution and square it.

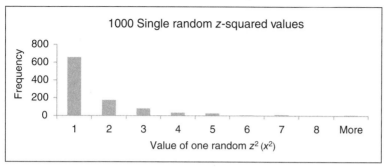

Figure 28.1 Remember that a given histogram bar is the number of results that are less than or equal to the bar's number label, and strictly greater than the number label of the bar to its left. So, for example, the 2 bar is the number of results that are ≤ 2 and > 1.

Figure 28.2 is a histogram of 1000 simulation results when we randomly select two values from the z-distribution and sum the two squared values.

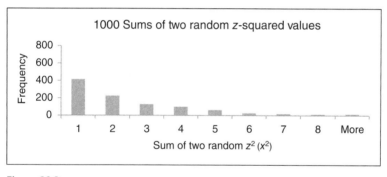

Figure 28.2

And finally, Figure 28.3 is a histogram of 1000 simulation results when we randomly select three values from the z-distribution and sum the three squared values.

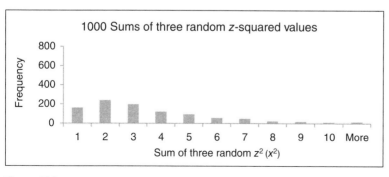

Figure 28.3

For the scenario we have $X^2 = 3.33 + 0 + 3.33 = 6.67$. Could this X^2 value have arisen by chance? Let's "zoom in" on the third histogram to see.

Looking at the histogram of Figure 28.4 the answer is: Yes, 6.67 can fairly easily arise by chance. It looks like about 100 out of the 1000 outcomes are above 7, so the p-value for 6.67 should be about 0.10. *Do not* reject the null hypothesis of a 30, 30, 30 split. The difference between actual and expected *is not* big enough.

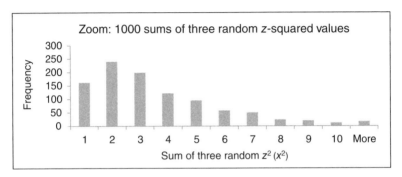

Figure 28.4

Notice that we are only concerned with the one-sided question of whether X^2 is large enough to reject the null hypothesis; analyzing X^2 is naturally a one-tail analysis. (See Chapter 16 if needed.)

Next let's zoom in on the histogram that is the sum of two random z-squared values. With the histogram of Figure 28.5, it seems the answer is: No, 6.67 is fairly unlikely to arise by chance, although it is a close call. Clearly less than 50 outcomes are above 7, so the p-value for 6.67 will probably be less than 0.05. *Do* reject the null hypothesis of 30, 30, 30 split. The difference between actual and expected *is* big enough.

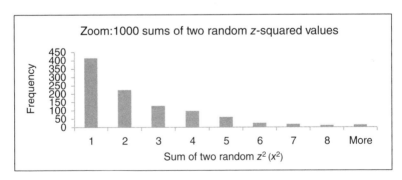

Figure 28.5

Figure 28.6 is a graph showing the three corresponding continuous probability distributions superimposed together. These are three separate Chi-squared (X^2) distributions.[2]

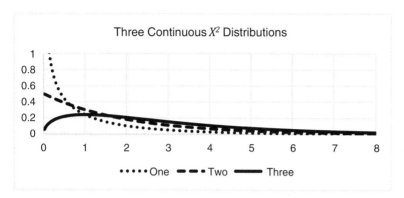

Figure 28.6

Which should we use? Here is a situation where "degrees of freedom" matters quite a bit.

---

2 The X^2 distribution becomes more and more normal. You can see the formula here: http://itl.nist.gov/div898/handbook/eda/section3/eda3666.htm

29

Facing Degrees of Freedom

> **Statistical Scenario—Degrees of Freedom**
>
> Can you figure out the value of x? You don't need to do any calculation, just answer Yes or No.
>
> $$2 + 4 + 3 + 8 + 1 + x = 0$$
>
> Answer: Yes.
>
> The value of x is dictated by the other values; x has no freedom.

I have mentioned "degrees of freedom" but I wrote that we would address it later. Later is now. Below is part of what I wrote while explaining the basics of the sample variance statistic:

To get the variance, we do the following . . . (don't worry about why 1 is subtracted from the sample size for now.)
The symbol for the sample variance is s^2.
Here is the sample variance calculation in math shorthand.

$$s^2 = \frac{1}{n-1} \sum_{i=1}^{n} (x_i - \bar{x})^2$$

Notice again that the formula for sample variance has $n - 1$ as a denominator. This term is the <u>degrees of freedom</u>.[1] In the above formula, the degrees of freedom are $n - 1$. 1 is subtracted from n because the formula uses the sample mean, which itself uses all the n data values. Question: If we know the value of the sample mean and all the data values except one, can we figure out the value of the remaining data value? Answer: Yes. The sum of all the $x_i - \bar{x}$ (unsquared)

---

1 Degrees of freedom is symbolized by *df* or ν (Greek lower case letter Nu), but I'll continue to spell it out.

Illuminating Statistical Analysis Using Scenarios and Simulations, First Edition.
Jeffrey E Kottemann.
© 2017 John Wiley & Sons, Inc. Published 2017 by John Wiley & Sons, Inc.

has to equal zero. So, one data value has no freedom when calculating s^2, hence $n - 1$ degrees of freedom.

With X^2 analysis, we have a similar situation, but the degrees of freedom for X^2 is dictated by the *number of categories*. Question: If we know all the expected (E) counts—we do because we specify them—and all the actual observed (O) counts except one, can we figure out the remaining actual count? Answer: Yes. The sum of all the $O_i - E_i$ (unsquared) has to equal zero. So, one count has no freedom when calculating X^2. In our current example, there are 3 categories D, R, and I. So, there are $3 - 1 = 2$ degrees of freedom. The degrees of freedom dictates which Chi-squared distribution to use.

Statistical analysis software automatically determines degrees of freedom for you.

30

Multiple Proportions: Goodness of Fit

Rather than eyeballing a histogram or a continuous probability distribution, let's look at the X^2 Table 30.1 that tells us the X^2 value that corresponds to various p-values and various degrees of freedom. Our X^2 value of 6.67 with two degrees of freedom corresponds to a p-value between 0.05 and 0.025. (A statistics calculator gives us the more precise p-value of 0.0356.) Notice that if there were three degrees of freedom, then the p-value would be between 0.10 and 0.05. These p-values mirror what we determined earlier with the simulation histograms.

Using an alpha-level of 0.05, our p-value of 0.0356 for X^2 of 6.67 with two degrees of freedom is considered statistically significant. Using an alpha-level of 0.01, it is not. (If you were reporting these results, you would report your alpha-level and whether the results are statistically significant given your alpha-level, but you would also need to report the p-value to let others evaluate the results using an alpha-level they might wish to use.)

The above example hypothesizes equal counts in each of the three categories. This need not be the case with the X^2 test. You can specify any counts in the various categories for the null hypothesis (of course they must sum to the total sample size). It is wonderfully flexible that way.

For example, let's say in a random sample of 120 you want to check whether there are equal numbers of D and R but twice as many I, which would be 30, 30, and 60. Say your sample gives you 28, 33, and 59. You would evaluate

$$X^2 = (28 - 30)^2/30 + (33 - 30)^2/30 + (59 - 60)^2/60 = 0.45$$

This X^2 of 0.45 is quite small and to the far left of the distribution. The p-value therefore will be large. The X^2 Table 30.1 indicates that the p-value will be greater than 0.25. A statistics calculator indicates that the p-value equals 0.79852 for X^2 of 0.45 with two degrees of freedom. Therefore, we will *not*

Illuminating Statistical Analysis Using Scenarios and Simulations, First Edition.
Jeffrey E Kottemann.
© 2017 John Wiley & Sons, Inc. Published 2017 by John Wiley & Sons, Inc.

Table 30.1 Chi-squared table.

X^2								p-Value								
		0.25	0.20	0.15	0.10	0.05	0.025	0.02	0.01	0.005	0.0025	0.001	0.0005
df	1	1.32	1.64	2.07	2.71	3.84	5.02	5.41	6.63	7.88	9.14	10.83	12.12
	2	2.77	3.22	3.79	4.61	**5.99**	**7.38**	7.82	9.21	10.60	11.98	13.82	15.20
	3	4.11	4.64	5.32	6.25	7.81	9.35	9.84	11.34	12.84	14.32	16.27	17.73

reject the null hypothesis of 30, 30, and 60. Keep in mind that we do not officially accept the null hypothesis, we just have not ruled it out.

> The <u>Chi-squared (X^2) goodness of fit test</u> is used to assess distributions of counts across categories.

As for minimum sample size requirements, a common rule of thumb is to have at least 20 overall and at least 5 in each category. With small samples, you can use what are called "exact tests"; do an online search for more on this.

A Note on Using Chi-squared to Test the Distribution of a Scaled Variable

(This can be skipped without loss of continuity.)

> *Statistical Scenario—Testing Scaled Variable Distributions*
>
> We have a random sample of 1000 male heights with a sample mean of 70 inches, sample variance of 9 inches, and sample standard deviation of 3. We want to test whether male heights approximate a normal distribution (the null hypothesis).

With a normal distribution, we expect 950 of the 1000 heights to be within the interval of 1.96 standard deviations above and below the mean, with 25 below that interval and 25 above that interval. The corresponding interval for our sample is $70 \pm 1.96 \times 3$, which is 64–76 inches. Looking at our data, let's say we find that 960 heights fall in this interval, with 24 below and 16 above.

The test: The null hypothesis expects the counts to be 25, 950, and 25; we have 24, 960, and 16 in our sample. Degrees of freedom is two. Plugging these numbers into a calculator, we get Chi-squared of 3.385 and p-value of 0.184. Using an alpha-level of 0.05, we do not reject the null hypothesis that male heights are normally distributed.

We could use more categories to capture finer intervals of the normal distribution, as shown in Figure 30.1. We would then compare the expected counts of 25, 135, 340, 340, 135, and 25 with the actual counts for these standard deviation intervals that we find in our sample of 1000 heights. (As you might imagine, if we make a large number of narrow intervals, we will almost always end up rejecting the null hypothesis of normality, even for data distributions that are quite close to normal.)

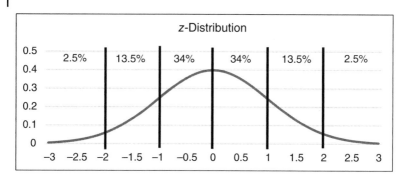

Figure 30.1

31

Two-Way Proportions: Homogeneity

A survey was conducted in the Flowing Wells (FW) community in which 100 random members recorded their political affiliation. The same survey was conducted in the Artesian Wells (AW) community asking 100 random members their political affiliation. Below are the observed counts of Democrats, Republicans, and Independents for FW and AW.

FW: 40, 30, and 30.
AW: 25, 30, and 45.

We want to assess whether the two communities might have different population proportions of Democrats, Republicans, and Independents.

Null hypothesis: No difference between the FW and AW communities' population proportions.

Rephrasing: How likely is it to get these sample counts if in fact FW and the AW have the same population proportions?

This scenario is an extension of the scenario in Chapter 12 in which we wanted to know if there was a proportion difference between FW and AW for a binomial variable. Now we are concerned with a multinomial variable.

To analyze this scenario, we can use the Chi-squared (X^2) Test for Homogeneity. A Chi-squared calculator tells us that the Chi-squared statistic is 6.462, degrees of freedom is two, and the p-value is 0.0395. Reject the null hypothesis that the two communities have the same population proportions of Democrats, Republicans, and Independents.

Detailed below is the arithmetic that yields the Chi-square of 6.462.

Table 31.1 shows the counts and totals we are dealing with.

We need to calculate what the expected counts are if the null hypothesis is true. Table 31.2 shows that. Each of the six cells equals its column total times its

Illuminating Statistical Analysis Using Scenarios and Simulations, First Edition.
Jeffrey E Kottemann.
© 2017 John Wiley & Sons, Inc. Published 2017 by John Wiley & Sons, Inc.

Table 31.1 Observed counts.

	Republican	Democrat	Independent	Totals
FW	40	30	30	100
AW	25	30	45	100
Totals	65	60	75	200

Table 31.2 Expected counts.

	Republican	Democrat	Independent	Totals
FW	$(65 \times 100)/200 = 32.5$	$(60 \times 100)/200 = 30$	$(75 \times 100)/200 = 37.5$	100
AW	$(65 \times 100)/200 = 32.5$	$(60 \times 100)/200 = 30$	$(75 \times 100)/200 = 37.5$	100
Totals	65	60	75	200

row total divided by the grand total of 200. Notice that all the totals remain as they were.

Table 31.3 shows each Chi-squared term, calculated as we have before.

$$\sum \frac{(\text{observed-expected})^2}{\text{expected}} = \text{the } \chi^2 \text{ statistic}$$

Summing up the six terms yields a Chi-square of 6.462. Calculating degrees of freedom is left for the next chapter.

Table 31.3 Chi-squared arithmetic terms.

	Republican	Democrat	Independent
FW	$(40 - 32.5)^2/32.5$	$(30 - 30)^2/30$	$(30 - 37.5)^2/37.5$
AW	$(25 - 32.5)^2/32.5$	$(30 - 30)^2/30$	$(45 - 37.5)^2/37.5$

32

Two-Way Proportions: Independence

> **Statistical Scenarios—College Majors**
>
> You have a random sample of local college student records that indicate student's sex and academic major. You want to know whether, in the campus population, certain majors are equally appealing to both sexes and whether certain majors appeal to one sex more than the other.
>
> Tables 32.1a and b have count breakdowns by sex and major. Which of the two tables seem to indicate that students' relative preference for the two majors is *independent* of their sex, and which of the two tables seem to indicate that students' relative preference for the two majors *depends* on their sex?

Table 32.1a&b

	Major1	Major2
Male	40	60
Female	18	32

	Major3	Major4
Male	90	10
Female	10	40

These two tables[1] portray scenarios that can be made obvious to the naked eye. Let's fill in some totals and percentages to make the relative proportions for males and for females easier to see—Tables 32.2a and b.

---

1 Tables like this are often called contingency tables or crosstabs.

Illuminating Statistical Analysis Using Scenarios and Simulations, First Edition.
Jeffrey E Kottemann.
© 2017 John Wiley & Sons, Inc. Published 2017 by John Wiley & Sons, Inc.

Table 32.2a&b

	Major1	Major2	Totals
Male	40 (40%)	60 (60%)	100 (100%)
Female	18 (36%)	32 (64%)	50 (100%)

	Major3	Major4	Totals
Male	90 (90%)	10 (10%)	100 (100%)
Female	10 (20%)	40 (80%)	50 (100%)

In Table 32.2a, we can see that similar proportions of males and females are enrolled in the two majors. Relative preference for each of the two majors seems independent of students' sex. To analyze this, we can use the Chi-squared (X^2) test for independence. (The arithmetic involved is the same as the test for homogeneity!) This scenario yields a low X^2 value and thus a high p-value: X^2 is 0.22 giving a p-value of 0.64. Verdict: Do not reject the null hypothesis of independence in the population between sex (Male, Female) and major (Major1, Major2).

Here is a useful way to think about independence: If you were asked to determine the approximate odds that an anonymous student was Major1 or Major2, would it help you to know the anonymous student's sex? Apparently not. The odds for males and females given our sample are quite similar: about 4–6 for both sexes.

In Table 32.2b, on the other hand, quite dissimilar proportions of males and females are enrolled in the two majors. Relative preference for each of the two majors seems to depend on students' sex. This scenario yields a high X^2 value and thus a low p-value: X^2 is 73.5 giving a p-value of less than 0.0001. Verdict: Reject the null hypothesis of independence in the population between sex (Male, Female) and major (Major3, Major4).

Here is a useful way to think about dependence: If you were asked to determine the approximate odds that an anonymous student was Major3 or Major4, would it help you to know the anonymous student's sex? Apparently yes. The odds for males and females given our sample are quite dissimilar: 9–1 for males and 1–4 for females.

The degrees of freedom for two-dimensional tables is the number of rows minus 1, times the number of columns minus 1. For the current scenarios that is $(2 - 1) \times (2 - 1) = 1$ degree of freedom. To see why, look at Table 32.3. You can freely pick a value between 0 and 58 for x in the upper left cell, but once you do, the values in the other three cells are determined. So, there is only one degree of freedom. (If you set x equal to 40, you'll get Table 31.1a.)

Table 32.3 One degree of freedom.

	Major1	Major2	Totals
Male	x	$100 - x$	100
Female	$58 - x$	$50 - (58 - x)$	50
Totals	58	92	150

Chi-squared calculators and statistical analysis software will determine degrees of freedom for you, and they will do all the Chi-squared arithmetic too.

Another Statistical Scenario—Sex and Ice Cream
Eyeballing Table 32.4, does there seem to be independence in the population between sex and the preferred ice cream flavor?

Table 32.4

	Vanilla	Chocolate	Strawberry	Totals
Male	25 (26%)	20 (21%)	50 (53%)	95 (100%)
Female	50 (34%)	75 (52%)	20 (14%)	145 (100%)

The Chi-squared test for independence can be performed with multinomial and not just binominal variables. The null hypothesis for this scenario is that preferred ice cream flavor is independent of sex. Looking at the table, however, there does not seem to be independence because the relative preferences appear to depend on whether someone is male or female.

The Chi-squared test for independence will give you a p-value to decide whether to reject the null hypothesis of independence in the population between two nominal variables that may be binomial or multinomial.[2] Using an online Chi-squared calculator, the X^2 for this scenario is 44.55 with $(2 - 1) \times (3 - 1) = 2$ degrees of freedom, corresponding to a p-value of less than 0.0001. Verdict: Reject the null hypothesis of independence in the population between sex and preferred ice cream flavor.

A word of caution is in order. To illustrate, let's combine the two tables from the original scenario into one table and analyze it—Table 32.5.

---

2 The variables involved are sometimes referred to as "factors."

Table 32.5 Combining Tables 32.1a and b.

	Major1	Major2	Major3	Major4	Totals
Male	40 (20%)	60 (30%)	90 (45%)	10 (05%)	200 (100%)
Female	18 (18%)	32 (32%)	10 (10%)	40 (40%)	100 (100%)

The X^2 for this combined scenario is 73.73 with $(2-1) \times (4-1) = 3$ degrees of freedom, corresponding to a p-value of less than 0.0001. Verdict: Reject the null hypothesis of independence between sex and major. Here, the similarity for the pair Majors 1 and 2 is "overwhelmed" by the dissimilarity for the pair Majors 3 and 4. The X^2 analysis above does not address such finer distinctions and simply suggests that dissimilarity lurks *somewhere*. Part VI Chapter 58 problem C highlights issues in "drilling down" to look at individual pairwise comparisons.

33

Variance Ratios and the *F*-Distribution

Statistical Scenario—Variance Difference

Random samples of 30 female and 30 male community members were surveyed for their opinions about a new policy using the following scale.

Strongly agree   1   2   3   4   5   6   7   Strongly disagree

You want to know whether the evidence suggests that there is a difference in the *variety* of opinions among the community's males and females.

We will use the *sample variances* as the sample statistic of variety.

Which of the survey sample variance ratio values in Table 33.1 convince you that there is a difference between the female and male *population variances*?

Said another way: How likely is it that you would get values for the sample variance ratio value as large as each of those shown in the table when randomly sampling from two populations that have equal population variances?

Null hypothesis: There is no difference in the variance of community opinion between males and females. That is, the population variance ratio equals 1.

Table 33.1 Scenario cases.

Female sample variance	Males sample variance	Sample variance ratio value
3	1	3
2	1	2
3	2	1.5
2	2	1

Illuminating Statistical Analysis Using Scenarios and Simulations, First Edition.
Jeffrey E Kottemann.
© 2017 John Wiley & Sons, Inc. Published 2017 by John Wiley & Sons, Inc.

To compare two sample variances, we need to (re)invent a sample statistic and the corresponding probability distribution. To make a sample statistic we can simply take the <u>ratio of two sample variances</u> and see if it is much different than 1, since a ratio value close to 1 suggests that the population variances may well be equal. We'll denote this ratio F.

Now we need a corresponding distribution for chance occurrences of F, called an F-<u>distribution</u>. Let's use simulation to construct it. As usual, we'll be simulating what to expect when the null hypothesis is true. Figure 33.1 is a histogram of sample variance ratios using sample sizes of 30 and 30 and equal population variances, making the population variance ratio equal to 1. To get a smooth and dense histogram, we'll simulate 10,000 surveyors, where each surveys 30 random males and 30 random females. Each surveyor calculates the sample variance for each of their two samples of 30 and calculates their sample variance ratio, F. Unbeknownst to the surveyors, males and females have the same population variance. Figure 33.1 is a histogram of the F values obtained by the 10,000 surveyors.

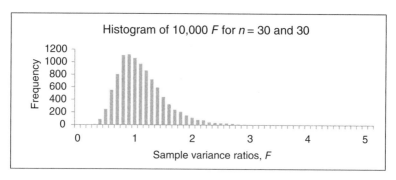

Figure 33.1

Traditionally, evaluations of F are one-tail. By making sure to put the larger of the two sample variances in the numerator of the ratio, we can focus on the right tail. Eyeballing this histogram you can see that sample variance ratios greater than 2 seem very unlikely to occur by chance when sampling 30 from each of two populations that have the same population variance. Less than 5% of the contents of this histogram are in the area from two to infinity, equating to a p-value of less than 0.05 (0.10 two-tail). So, we can be at least 95% confident that a sample variance ratio of 2 or more using sample sizes 30 and 30 will not come from a community in which males and females have equal population variances—see Table 33.2.

Table 33.2 Scenario solution.

Sample variance ratio value	
3	
2	Here and above, reject the null hypothesis when $n = 30$ and 30
1.5	
1	

What about when the sample sizes are 5 and 5?

What about 100 and 100?

It may come as no surprise that if our sample sizes are smaller than 30 and 30 then the variance ratio (F) cutoff will have to be larger than 2. The less evidence you have, the more conclusive that evidence needs to be. Figure 33.2 shows simulation results for sample sizes 5 and 5, where both samples are drawn from populations that have the same population variance. You can see that a substantial percentage of the sample variance ratios now exceed 2 simply by chance. (The rightmost bar is for all F greater than 5.) About 25% (2,500/10,000) of the sample variance ratios are greater than 2, equating to a p-value of about 0.25. To get a p-value of 0.05, it looks like F will need to be somewhere around 6 since the p-value of F greater than 5 is about 0.07 (the rightmost bar has about 700 of the 10,000 results).

And when sample sizes are 100 and 100 as shown in Figure 33.3, almost no F ratios are greater than 2, equating to a very small p-value. To get a p-value of 0.05, it looks like F will need to be about 1.5.

Figure 33.4 is a graph showing the three corresponding continuous probability distributions superimposed together. These are three separate F-distributions.[1]

---

1 You can see the formula here: http://itl.nist.gov/div898/handbook/eda/section3/eda3665.htm. Sample variances are distributed in the shape of the X^2 distribution. You will see this if you compare the shapes of the X^2 histograms of Chapter 28 with those of s^2 in Chapter 27; both involve sums of squared differences (deviations). The ratio of two variances is therefore distributed in accordance with the ratio of two X^2 distributions: That is what the F-distribution is. Notice that the F-distribution becomes more and more normal.

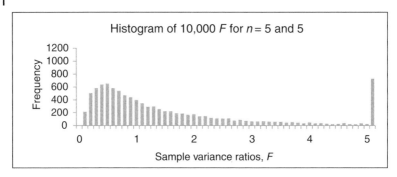

Figure 33.2

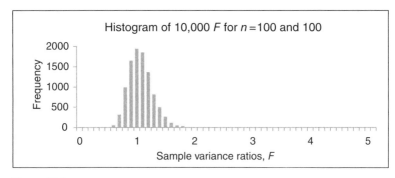

Figure 33.3

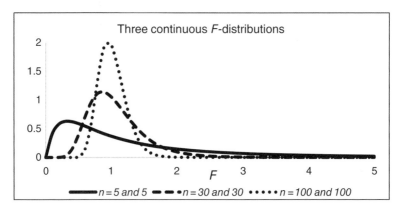

Figure 33.4

Mirroring what we did with the histograms above, we can determine p-values for any F value using an F-test calculator. For an F value of 2, the p-value for n of 5 and 5 is 0.2593, the p-value for n of 30 and 30 is 0.0334, and the p-value for n of 100 and 100 is 0.0003.

Also mirroring what we did with the histograms above, we can also determine any alpha-level cutoff for F using the calculator. The 0.05 alpha-level cutoffs for F are 6.39 for n of 5 and 5, 1.86 for n of 30 and 30, and 1.39 for n of 100 and 100.

Statistical analysis software usually determines the appropriate degrees of freedom for you, but online calculators often do not. For the statistical analyses in this chapter, the degrees of freedom for the numerator and the denominator are the respective sample sizes minus one. These two distinct degrees of freedom dictate which particular F-distribution is applicable.

> F-distributions are used to analyze sample variance ratios, F.

34

Multiple Means and Variance Ratios: ANOVA

Statistical Scenarios—Analysis of Variance (ANOVA)

You have 1–7 scaled opinions for random samples of 30 Democrats, 30 Republicans, and 30 Independents in the community. You want to know whether the evidence suggests that there is a difference in the *average opinion* between *any* of the three political groups in the overall community population. *Null hypothesis*: there is not.

For each of the four separate scenarios below, which sample statistic values convince you that there is a population difference in the means between at least two of the political groups? Pay attention to (1) how far apart the sample means are, and (2) how high the sample variances are.

Scenario 1

Opinion statistics	Democrat	Republican	Independent	
Sample mean	3.9	4.0	4.1	Means close together
Sample variance	0.9	1.0	1.1	Low variance

Scenario 2

Opinion statistics	Democrat	Republican	Independent	
Sample mean	3.0	4.0	5.0	Means Far apart
Sample variance	0.9	1.0	1.1	Low variance

Scenario 3

Opinion statistics	Democrat	Republican	Independent	
Sample mean	3.9	4.0	5.0	One mean far apart
Sample variance	0.9	1.0	1.1	Low variance

Illuminating Statistical Analysis Using Scenarios and Simulations, First Edition.
Jeffrey E Kottemann.
© 2017 John Wiley & Sons, Inc. Published 2017 by John Wiley & Sons, Inc.

Scenario 4

Opinion statistics	Democrat	Republican	Independent	
Sample mean	3.0	4.0	5.0	Means far apart
Sample variance	8.9	9.0	9.1	Higher variance

Variance ratios and the *F*-distribution can be used in an ingenious way to *broadly* analyze sample mean differences. The analysis method we'll be looking at in this chapter and the next—analysis of variance (ANOVA)—does not reveal which specific groups are different from which, just whether at least two groups' samples may represent populations with different population means. However, as we'll see, ANOVA needs to assume that all samples come from populations with the same population variance.

If all the groups' samples in a given scenario come from populations with the same population mean (the null hypothesis) and population variance (an ANOVA assumption), then all the groups' sample means will be members of a common sample mean distribution. For example, Figure 34.1 is a sampling simulation for a population with population mean 4 and population variance 1. We'll use it to assess the first three scenarios. Note that the average of the sample variances for each of these three scenarios equals 1, which is used for the simulation.

For the first scenario, notice that all the sample means (3.9, 4, and 4.1) are clearly members of this one sampling distribution histogram. This suggests that we should not reject the null hypothesis of equal population means. On the other hand, for the sample means of the second scenario (3, 4, and 5), the body of the sampling distribution histogram contains 4, but 3 is far to the left and 5 is far to the right. This suggests that we should reject the null hypothesis of equal population means; the population variance would have to be much larger than 1 for the sampling distribution to become wide enough to contain 3, 4, and 5 (see Chapter 21 if needed). For the third scenario with sample means of 3.9, 4, and 5, both 3.9 and 4 are members, but 5 is far to the right. This also suggests that we

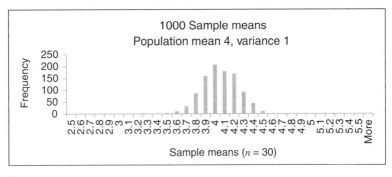

Figure 34.1

should reject the null hypothesis of equal population means; the population variance would have to be larger than 1 for the sampling distribution to become wide enough to contain 3.9, 4, and 5.

Next, let's look at how we can conduct a similar analysis formulaically. We'll start with analyzing the second scenario: sample means 3, 4, and 5, sample variances equal 0.9, 1.0, and 1.1, and all sample sizes equal 30.

First, we'll use the sample variances to estimate the population variance, σ^2. Since we are assuming that all three samples are drawn from populations with the same population variance (an ANOVA assumption), and since the sample sizes are the same for all three, we can simply take the average of our sample variances to get a *direct estimate* of the common population variance: $(0.9 + 1.0 + 1.1)/3$ equals 1.[1]

Second, we want to determine whether a population with variance equal to 1 results in random samples that have sample means as far apart as 3, 4, and 5. Determining this requires several steps.

Recall from Chapter 22 that with a sample mean distribution we know that $SE = \sqrt{\sigma^2/n}$, where SE is standard error, σ^2 is population variance, and n is sample size. We can rearrange this formula by squaring both sides to get $SE^2 = \sigma^2/n$ and then multiplying both sides by n to get

$$SE^2 \times n = \sigma^2$$

SE^2 is the variance of the sample means distribution (the square of standard error).[2]

We can estimate SE^2 for the scenario by calculating the variance of our actual sample means. The average of the sample means 3, 4, and 5 equals 4, so the variance of the sample means equals $(3 - 4)^2 + (4 - 4)^2 + (5 - 4)^2$ divided by 2 degrees of freedom equals 1.

Now we can calculate an *indirect estimate* for σ^2 using $SE^2 \times n$ giving $1 \times 30 = 30$.

This implies that the sample means 3, 4, and 5 will be members of a common sample mean distribution if the population variance is 30 (or more). See Figure 34.2 for a histogram of this hypothetical situation.

But the *direct* estimate we got for the scenario's population variance is only 1, which is too small to accommodate the sample means of 3, 4, and 5. (See the histogram in Figure 34.1 and recall that smaller population variances have narrower sample mean distributions.)

---

1 If the sample sizes are not equal, we can introduce corresponding weightings into the equations. *P*-value calculators and statistical analysis software will do this for you.
2 If you have not already, you may want to look at Appendix C: Standard Error as Standard Deviation.

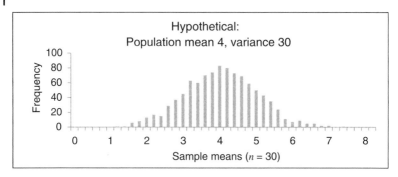

Figure 34.2

To formally compare the indirect and direct variance estimates, we make a ratio of them:

$$\frac{\text{variance given the distance } between \text{ the groups' sample means (the indirect estimate of } \sigma^2)}{\text{variance given the average variance } within \text{ the groups (the direct estimate of } \sigma^2)}$$

For the second scenario, the variance ratio F is $30/1$.

When the numerator is greater than the denominator to a degree that is unlikely to happen by chance, we will reject the null hypothesis of equal population means. Since we want to get a p-value for a variance ratio, we use the F-distribution. And, since we are only interested in whether F is large enough, the p-value of interest is one-tail.

Degrees of freedom for the numerator is the number of categories minus 1, as we saw with Chi-squared. Degrees of freedom for the denominator is the total sample size minus the number of categories, which is equivalent to summing $n - 1$ across all the categories. These two distinct degrees of freedom dictate which particular F-distribution is applicable; for all four scenarios, with three categories and 30 observations per category, the applicable F-distribution is symbolized by $F(2,87)$ and is shown in Figure 34.3.

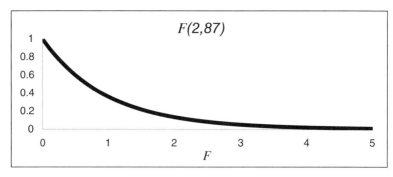

Figure 34.3

For the second scenario, the variance ratio F of $30/1 = 30$ with 2 and 87 degrees of freedom yields a p-value of less than 0.00001. (This is way off the chart shown in Figure 34.3) Reject the null hypothesis of equal population means.

The first scenario with sample means of 3.9, 4, and 4.1 gives a much smaller estimate for SE^2 of $(3.9 - 4)^2 + (4 - 4)^2 + (4.1 - 4)^2$ divided by 2 degrees of freedom equaling 0.01. The indirect estimate for the population variance $SE^2 \times n$ is therefore also much smaller at 0.01×30 equaling 0.3. The direct estimate for σ^2 is the average of the sample variances and equals 1. F of $0.3/1 = 0.3$ with 2 and 87 degrees of freedom yields a p-value of 0.74. Do not reject the null hypothesis of equal population means.[3]

Next, let's look at the fourth scenario with sample means of 3, 4, and 5, sample variances equal 8.9, 9.0, and 9.1, and all sample sizes equal 30. Figure 34.4 shows the results of a sampling simulation. (Recall that larger population variances have wider sample mean distributions.)

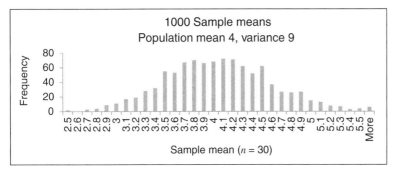

Figure 34.4

For this scenario, the sample mean of 4 is obviously in the sampling distribution, but 3 and 5 are located out in the tails. Therefore, it is fairly unlikely that they represent populations with the same population mean.

Formulaically, the sample means of 3, 4, and 5 give an estimate for SE^2 of 1. $SE^2 \times n$ is 1×30 equaling 30. The direct estimate for σ^2 is the average of the sample variances and equals 9. F of $30/9 = 3.33$ with 2 and 87 degrees of

---

3 In general, the variance ratio gets smaller for any number of reasons: When means are closer together, because that decreases the SE^2 estimate in the numerator; when n decreases, given its presence as a multiplier in the numerator; and when sample variances are higher, because that increases the denominator. Note that the denominator is not affected by how far apart the sample means are.

freedom yields a *p*-value of approximately 0.05. Therefore, it is fairly unlikely that they represent populations with the same population mean.

You can work out the details for the third scenario, reproduced below. (You should get F of $11.1/1 = 11.1$ giving a *p*-value of approximately 0.0003)

Scenario 3

Opinion statistics	Democrat	Republican	Independent
Sample mean	3.9	4.0	5.0
Sample variance	0.9	1.0	1.1

As noted, ANOVA does not reveal which specific groups are different from which, just whether at least two groups' samples may represent populations with different population means. In scenario 3 it appears that the difference between the sample mean pairs of Democrat versus Independent and Republican versus Independent dominate the ANOVA results. Part VI Chapter 58 problem E highlights issues in "drilling down" to look at individual pairwise comparisons.

To conclude this chapter, Figure 34.5 shows where each of the four scenarios maps onto the *F(2,87)* distribution.

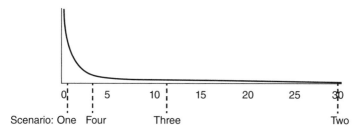

Scenario: One Four   Three   Two

Figure 34.5

35

Two-Way Means and Variance Ratios: ANOVA

The previous chapter dealt with one-way ANOVA, also called single-factor ANOVA. Two-way ANOVA is a powerful extension that allows you to assess scaled data across two factors (two nominal variables). Further, as we'll see, it detects possible interactions between the two factors. The two factors can have any number of categories, but we'll look at the simplest case where each factor has only two categories. The basic layout is shown in Table 35.1. The inner four cells give the sample means for the subgroups of DM, DF, RM and RF.

Say, that a survey was conducted and respondents indicated how closely they follow politics on a 1–7 scale. Overall differences across political affiliations (D&R) can be assessed using the rightmost column and the overall differences across sexes (M&F) can be assessed using the bottom row. These are called the two main effects. Further, we can detect possible interaction effects between the two factors by looking for patterns in the inner cells. In particular, we can look to see whether DM mean minus DF mean is different from RM mean minus RF mean.

Study Tables 35.2–35.7, which are hypothetical scenarios to clearly show you the various types of statistical situations. Assume the sample sizes are large enough and the sample variances are small enough so that the DM, DF, RM and RF sample means are very reliable estimates of their respective population means.

The arithmetic involved is fairly tedious and won't be covered here. ANOVA calculators and statistical analysis software will do it for you and give you p-values for each of the main effects as well for interaction effects.

More complex situations can also be analyzed with ANOVA, or one of its variants. Further, as we'll see midway through Part IV, there are other ways to conduct such analyses.

> This marks the end of Part III. At this point you can look at the review and additional concepts in Part VI Chapter 58, or proceed directly to Part IV.

Illuminating Statistical Analysis Using Scenarios and Simulations, First Edition.
Jeffrey E Kottemann.
© 2017 John Wiley & Sons, Inc. Published 2017 by John Wiley & Sons, Inc.

Table 35.1 Scenario layout.

	Males	Females	
Democrat	DM mean	DF mean	Overall D mean
Republican	RM mean	RF mean	Overall R mean
	Overall M mean	Overall F mean	Grand mean

Table 35.2 Scenario A.

	Males	Females	
Democrat	4	4	4
Republican	4	4	4
	4	4	4
No differences by sex (bottom row) or by political affiliation (right column).			
No interaction. $4 - 4 = 0$ and $4 - 4 = 0$			

Table 35.3 Scenario B.

	Males	Females	
Democrat	6	2	4
Republican	6	2	4
	6	2	4
Difference by sex (bottom row) but not by political affiliation (right column).			
No interaction. $6 - 2 = 4$ and $6 - 2 = 4$.			

Table 35.4 Scenario C.

	Males	Females	
Democrat	6	6	6
Republican	2	2	2
	4	4	4

Difference by political affiliation (right column) but not by sex (bottom row).

No interaction. $6 - 6 = 0$ and $2 - 2 = 0$.

Table 35.5 Scenario D.

	Males	Females	
Democrat	6	4	5
Republican	3	1	2
	4.5	2.5	3.5

Difference by both sex (bottom row) and political affiliation (right column).

No interaction. $6 - 4 = 2$ and $3 - 1 = 2$.

Table 35.6 Scenario E.

	Males	Females	
Democrat	6	2	4
Republican	2	6	4
	4	4	4

No differences by sex (bottom row) or by political affiliation (right column).

There is evidence of an interaction because $6 - 2 = 4$ but $2 - 6 = -4$.

Interaction effect: Male mean is higher when D, and female mean is higher when R.

Table 35.7 Scenario F.

	Males	Females	
Democrat	2	2	2
Republican	2	6	4
	2	4	3

While the bottom row and the right column indicate differences both by sex and by political affiliation, the real story is the interaction effect.

There is evidence of an interaction because $2 - 2 = 0$ but $2 - 6 = -4$.

Interaction effect: Female mean is higher when R.

Part IV
Linear Associations: Covariance, Correlation, and Regression

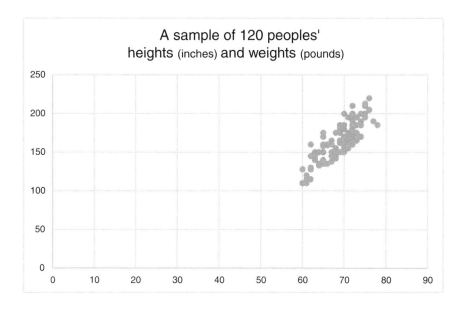

A sample of 120 peoples' heights (inches) and weights (pounds)

Illuminating Statistical Analysis Using Scenarios and Simulations, First Edition.
Jeffrey E Kottemann.
© 2017 John Wiley & Sons, Inc. Published 2017 by John Wiley & Sons, Inc.

36

Covariance

We now turn our attention to some of the most widely used methods in the statistical tool kit. By the end of Part IV we will have constructed a statistical model that predicts peoples' weight (scaled data) based upon their height (scaled data) and their sex (binomial data). We'll start off by looking at a visual aid called a scatterplot (Figure 36.1). It paints a pointillist picture of the relationship between pairs of variables (X, Y). We'll make X be peoples' heights and Y be peoples' weights. Each <u>data point</u> (x, y) on the scatterplot is the height and weight of one person.

Statistical Scenario—Height & Weight Scatterplot and Covariance
We have the heights and weights for a sample of 120 people.
We want to know whether height and weight are related.
In other words, does knowing someone's height help us guess their weight?
Null hypothesis: No (linear) relationship between height and weight.

Figure 36.1 is a scatterplot showing a sample of peoples' actual heights and weights. All the people are adults.[1] Lines showing the sample means for height ($\bar{x} = 69$) and weight ($\bar{y} = 164$) have been added to the scatterplot. Also included are the regions denoted by letters A, B, C, and D. In region A, height is below average and weight is above average; in B, both height and weight are above average. In C, height is above average and weight is below average. In D, both height and weight are below average.

How well do peoples' heights predict their weights?

---

1 The data can be found in Appendix D. Many "extreme" cases – short and heavy or tall and light – are omitted so that the relationship is more obvious. We will consider the full sample in Chapter 47.

Illuminating Statistical Analysis Using Scenarios and Simulations, First Edition.
Jeffrey E Kottemann.
© 2017 John Wiley & Sons, Inc. Published 2017 by John Wiley & Sons, Inc.

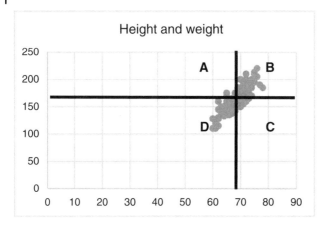

Figure 36.1

Before we try to answer that question, let's see what a scatterplot looks like when there is a perfect linear relationship between two variables X and Y. As shown in Figure 36.2, a perfect linear relationship is—no surprise—a perfectly straight line. If the relationship between height and weight were this perfect we could predict everyone's weight perfectly if we knew their height.

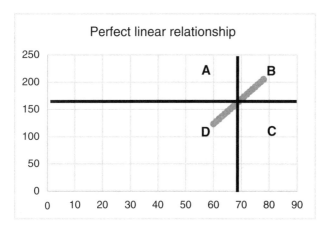

Figure 36.2

Now look at the scatterplot in Figure 36.3 where there is no discernable relationship between X and Y. The heights (X) are the actual heights, but the weights (Y) are just *random numbers*. This scatterplot is indistinguishable from random splatter. Knowing height does not help us predict weight.

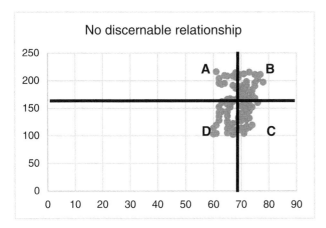

Figure 36.3

The scatterplot for actual heights and weights is somewhere between perfect and random. Knowing someone's weight will do us some degree of good in predicting their weight. But what specifically is "some degree of good?" We need a statistic that reflects how good the relationship is.

To (re)invent a linear relationship statistic, let's make some observations about the three scatterplots. When a perfectly positive linear relationship exists (Figure 36.2), if someone's height is above average then their weight will be above average too. This will always be true. And if someone's height is below average then their weight will be below average. This will always be true. No exceptions. The regions labeled B and D contain those cases. The exceptions will be in regions A and C, and there are none. In the perfect linear relationship all the points are in the regions B or D and none are in A or C.

In the no discernable relationship (Figure 36.3) this is not at all true. The A, B, C, and D regions all contain points, and lots of them. It is indistinguishable from random splatter.

The actual height and weight data (Figure 36.1) is somewhere in between: Noticeably more points, but not all, are in B and D rather than A and C. This is key to (re)inventing a statistic to reflect linear relationships. Since we are concerned with how two variables covary, let's call the statistic "covariance" like everyone else does. As we'll see, the covariance statistic is fundamental, and it will form the basis for other extremely useful statistics.

Now consider the points (x, y) in each of the four regions. *Question*: Which points correspond to negative products and which to positive products when we do the following bit of arithmetic?

(value of x – mean of x) × (value of y – mean of y)

Answer: The products for points in A and C are negative because negative multiplied by positive is negative. Those points are above one mean and below the other. The products for points in region B are positive because positive multiplied by positive is positive. Those points are above both means. The products for points in region D are also positive because negative times negative is positive. Those points are below both means.

If we add up the products for all the points, what do we get? In this case (Figure 36.1), we'll get a positive number. This indicates a positive relationship. Finally, we average the products using the degrees of freedom, which is the number of products minus 1 (since the sample means are used in the calculation, one of the products can not be considered free per Chapter 29). This gives us the sample covariance.

Here is the sample covariance s_{xy} calculation in math shorthand.

$$s_{xy} = \frac{1}{n-1} \sum_{i=1}^{n} (x_i - \bar{x}) \times (y_i - \bar{y})$$

Using the actual data, we get a sample covariance of 72.56.

The sample covariance of 72.56 suggests a positive linear relationship between height and weight, but interpreting the magnitude of the covariance is difficult. In the next chapter we'll look at a statistic that is much easier to interpret. In the chapter after that we'll determine its p-value.

Note: Always keep in mind that covariance—and the other statistics built from it—captures linear relationships. Figure 36.4 shows a nearly perfect nonlinear relationship, but it has near zero covariance. The covariance products in the A and B regions cancel each other out, as do the covariance products in the C and D regions.

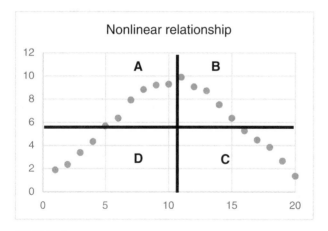

Figure 36.4

37

Correlation

Statistical Scenario—Height & Weight Correlation

Interpreting the magnitude of a covariance is difficult.

Is there a better statistic for interpretation?

Since covariance is made up of this

(value of x − mean of x) × (value of y − mean of y)

the unit of covariance is the unit of X times the unit of Y. This is not usually easy to interpret; interpreting the unit (height × weight) is not something we typically do. Luckily we can easily transform covariance into something that is a dream to interpret.

What we want is a standardized statistic, which we can make by rescaling the covariance of X and Y using the standard deviations of X and Y. The resulting statistic of "corelation" is called Pearson's correlation coefficient, symbolized by r, which tells us on a standardized −1.0 to +1.0 scale how closely two variables covary (and, like covariance, whether the relationship is positive or negative).

Here is the sample correlation calculation in math shorthand.

$$r = \frac{s_{xy}}{s_x \times s_y} = \frac{1}{n-1} \sum_{i=1}^{n} \frac{(x_i - \bar{x})}{s_x} \times \frac{(y_i - \bar{y})}{s_y}$$

When we divide the covariance (s_{xy}) by the product of the standard deviations of the two variables involved ($s_x \times s_y$) we'll get a value between −1.0 and +1.0 that is "unitless" because the numerator and denominator units cancel each other out. This is much easier to interpret.

Illuminating Statistical Analysis Using Scenarios and Simulations, First Edition.
Jeffrey E Kottemann.
© 2017 John Wiley & Sons, Inc. Published 2017 by John Wiley & Sons, Inc.

> With Pearson's correlation coefficient, r, -1.0 is a perfect negative linear relationship, 0.0 is the absence of a linear relationship, and $+1.0$ is a perfect positive linear relationship.

Using the actual data, we get a sample correlation $r = 0.81$, rounded to two decimal places.

The correlation of 0.81 is much closer to the perfect relationship value of 1.0 than it is to the no discernable relationship value of 0.0. Next, we'll determine the p-value.

38

What Correlations Happen Just by Chance?

Statistical Scenario—Sample Correlations Distribution
What will the distribution of sample correlations look like?
How do we determine p-values?

How likely is it that a given sample correlation coefficient value (0.81 in this case) could arise by chance? Let's do a simulation involving pure chance. We'll do 10,000 correlations using nothing but purely random numbers. We'll use the same sample size as the *statistical scenario* and correlate 120 pairs of random numbers 10,000 separate times. Figure 38.1 is the histogram of simulation results. Notice that it is centered on zero and has a normal bell shape. This shows us what to expect when sampling from a population with a *population correlation equal to zero*, which is the null hypothesis.[1]

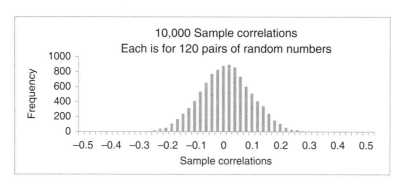

Figure 38.1

_____

1 The Greek letter rho is used to designate the *population* correlation, ρ_{XY}, but I'll continue to spell it out.

Illuminating Statistical Analysis Using Scenarios and Simulations, First Edition.
Jeffrey E Kottemann.

You can see that sample correlations less than −0.2 or greater than +0.2 are unlikely to happen by chance with a sample size of 120 number pairs. Our *statistical scenario* sample correlation of 0.81 is almost impossible by chance. Its *p*-value will be *extraordinarily* small.

What if we had a smaller sample size, say 30 number pairs? With smaller sample sizes, more extreme correlations should be more likely by chance. Figure 38.2 shows the simulation results of 10,000 correlations for samples of 30 pairs of random numbers.

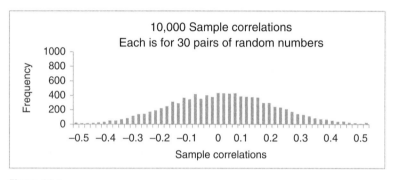

Figure 38.2

Sure enough, correlations of −0.2 and +0.2 are now fairly likely by chance. (A sample correlation of 0.81 is still almost impossible by chance, even with this small sample size.) As we'll see, the formulaic approach captures this sample size dynamic.

Below are the formulas for the <u>standard error of a sample correlation</u> and the number of standard errors for a sample correlation, #SEs. The #SEs formula rescales *r* in terms of standard error—the same principle we have seen before with sample proportions and sample means.

$$\text{Standard Error} = \sqrt{\frac{1 - r^2}{n - 2}}$$

$$\#\text{SEs} = \frac{r}{\text{Standard Error}}$$

Combining and rearranging terms we get

$$\#\text{SEs} = r \times \sqrt{\frac{n - 2}{1 - r^2}}$$

Let's interpret this formula for *n* and *r*. As *n* increases, #SEs will move farther from zero, and the *p*-value will decrease; more evidence leads to less uncertainty. Also, as *r* moves farther from zero, #SEs will move farther from zero, and the *p*-value will decrease; the further the correlation is from zero, the less likely it arose simply by chance.

> The *p*-value, and hence the statistical significance, for a sample correlation is determined by its magnitude and its sample size

Based on calculations using the formula, our correlation of 0.81 with sample size 120 equates to about 15 standard errors. That is, 15 standard errors out in the tail, which is way, way, way out in the tail. To get a *p*-value, we use the *t*-distribution. Using the *t*-distribution for 118 degrees of freedom, 15 standard errors from zero yields an extremely small *p*-value of 10^{-29} (0.00000000000000000000000000001).

Note: The *t*-distribution is used because sample (rather than population) standard deviations are used to calculate *r*, which makes standard error and #SEs only estimates of their true values (Chapter 27). And, we have $120 - 2 = 118$ degrees of freedom because the pair of sample means and the pair of sample standard deviations are used to calculate the sample correlation, so two of the data pairs can not be considered free (Chapter 29).

The correlation of 0.81 is strong and its *p*-value is extraordinarily small (10^{-29}). Reject the null hypothesis of no linear relationship between height and weight.

> Now we have a way to assess the strength (*r*) and the statistical significance (*p*-value) of a linear relationship between two variables.

> Statistics of strength, such as *r*, are often called statistics of <u>effect size</u>.[2] (As in, the "effect" that height has on weight.)

Special Considerations: Confidence Intervals for Sample Correlations

(This section can be skipped without loss of continuity.)

There is a complication that must be addressed by formulaic methods when determining confidence intervals for correlations: The sampling distributions for correlations become more compact and deform from normal symmetry as they approach the boundaries of −1 and +1. You can see this in Figure 38.3, which is a histogram of correlations for random samples of size 30 taken from a population in which the population correlation is 0.8.

---

2 Effect size statistics can also be calculated to assess the strength of results for other statistical tests, such as *t*-tests, Chi-squared tests, and ANOVA. For more, see Chapter 57 problem H, and do an online search for "effect size statistics."

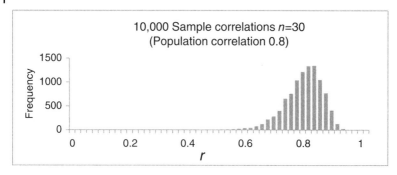

Figure 38.3

To get around this problem, formulaic methods for determining sample correlation confidence intervals will (1) convert correlations into special normally distributed z-values, symbolized by z', using Fisher's r to z' transformation formula, (2) calculate standard error and confidence intervals for the z' distribution, and then (3) convert the z' values back to correlations using the inverse z' to r transformation formula.

To help you visualize the relationship between r and z', Figure 38.4 shows the z' transform of Figure 38.3.

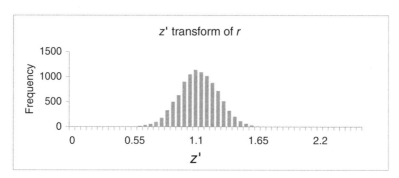

Figure 38.4

Here are the formulaic details:

The r to z' formula is $1/2\ln((1+r)/(1-r))$, where ln is the natural (base e) logarithm. The standard error of the z' sampling distribution is approximated by $(1/\sqrt{n-3})$. The inverse z' to r formula is $(e^{2z'}-1)/(e^{2z'}+1)$.

You can refer to Figures 38.3 and 38.4 as we overview the arithmetic: Using the r to z' formula with r of 0.8 gives us z' of approximately 1.1. The standard error formula gives us approximately 0.19. So, $1.1 \pm 1.96 \times 0.19$ gives us z' from 0.72 to 1.48 as the 95% confidence interval in terms of z'. Converting this interval back to r using the z' to r formula gives us 0.62–0.90 as the 95% confidence interval for our sample correlation. This agrees with the original simulation histogram in Figure 38.3, where we can see the confidence interval for r directly!

Statistics calculators will do all this for you. Using a calculator for our 0.81 sample correlation between height and weight for our sample of 120 people gives us a 95% confidence interval for r of 0.738–0.863 and a wider 99% confidence interval of 0.711–0.877. Figure 38.5 shows the sampling distribution for r around 0.81 with sample sizes of 120 constructed via simulation, and Figure 38.6 shows the distribution in terms of z'.

> The width of a sample correlation's confidence interval depends on its absolute magnitude as well as sample size.

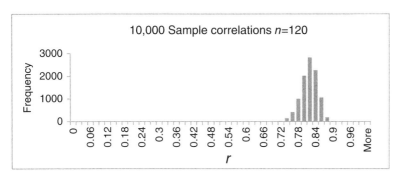

Figure 38.5

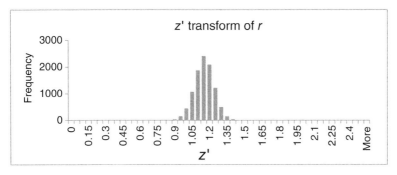

Figure 38.6

39

Judging Correlation Differences

Statistical Scenario—Correlation Differences

Is there a difference in the predictability of male versus female weights based on heights?

Null hypothesis: There is no difference between the population correlation of males' heights and weights and the population correlation of females' heights and weights.

With our actual height and weight data, the male sample correlation is 0.65 and the female sample correlation is 0.77, for a difference of −0.12. This difference suggests the *possibility* that females' weights are more predictable, but we need to determine how likely it is that such a difference in sample correlations could simply arise by chance.

In the following simulation, we'll assume the null hypothesis is true, with the same population correlation for males and females. For the population correlation we'll use the average of 0.65 and 0.77, which is 0.71.[1] Each of 10,000 surveyors gathers height and weight data for 60 random males and 60 random females, calculates the male and the female sample correlations, and then calculates the correlation difference of males minus females.[2] We want to see how many of the 10,000 differences are less than or equal to −0.12. To make it two-tail, we'll also consider differences greater than or equal to +0.12. This will show us how likely it is to get sample correlation differences as extreme as −0.12 or +0.12 when the null hypothesis is actually true. Figure 39.1 is the simulation results histogram.

---

1 The average is not precisely reflective due to *special considerations*, but is okay for our purposes.
2 There are more males than females in our actual sample, but for simplicity this chapter assumes equal numbers of each.

Illuminating Statistical Analysis Using Scenarios and Simulations, First Edition.
Jeffrey E Kottemann.
© 2017 John Wiley & Sons, Inc. Published 2017 by John Wiley & Sons, Inc.

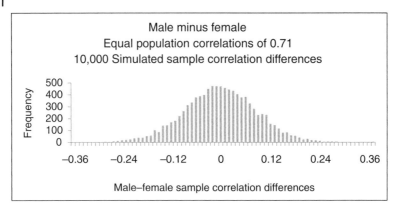

Figure 39.1

Approximately 20% of the simulation results are less than or equal to −0.12 or are greater than or equal to +0.12, indicating that our actual sample correlation difference is somewhat likely to occur when the null hypothesis is true. Using a *p*-value calculator for sample correlations of 0.65 and 0.77 with sample sizes of 60 and 60 yields a two-tail *p*-value of approximately 0.20.

Based on our evidence, we cannot convincingly claim that the population height and weight correlations for males and females are different. Relatedly, we can't say that there is a sex difference in (linear) predictive accuracy for weights based upon heights.

Special Considerations: Sample Correlation Differences

(This section can be skipped without loss of continuity.)

As we saw in the previous chapter's *special considerations*, sampling distribution histograms become more compact and deform from normal symmetry as they approach the boundaries of −1 or +1. What happens when we consider two correlation sampling distributions together? Compare the two multifaceted histograms in Figure 39.2a and b. Both involve random sampling from two different populations depicted on the same histogram, and in both cases the two different populations have underlying population correlations that are 0.2 apart. But, the case in Figure 39.2b involves a correlation that is closer to the boundary of +1.

Notice that the histogram bars for the first case span a width of about 0.6 (from 0.25 to 0.85) while the bars for the second case span a width of about 0.4 (from 0.55 to 0.95). Therefore, the distribution of sample correlation *differences* that we expect by chance should be more compact for the second case. You can see this in the simulation histograms of the differences between sample correlations shown in Figure 39.3a and b.

Figure 39.2

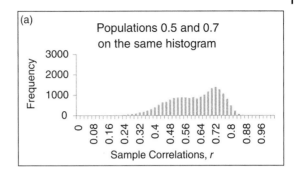

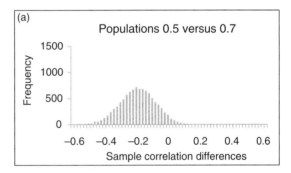

Figure 39.3

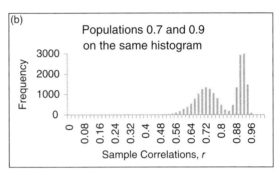

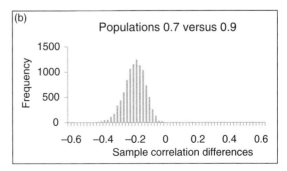

Also notice that as the population correlations get closer to the boundary of −1 or +1, the distribution of sample correlation differences will itself lose normal symmetry. To account for this, formulaic methods to compare two sample correlations will use Fisher transformations, similar to what we saw in the previous Chapter's *special considerations*.

The *p*-value, and hence the statistical significance, of the *difference* between two sample correlations is determined by the relative *and* the absolute magnitudes of the correlations, as well as the sample sizes.

40

Correlation with Mixed Data Types

Statistical Scenario—Correlation Between a Binomial and a Scaled Variable

We can also use the correlation approach we have been using to assess the relationship between a 0 or 1 valued binomial, such as sex, and a scaled variable, such as height.

For sex and height, calculations give us r of -0.67 and an extremely small p-value of 10^{-16}. We reject the null hypothesis of no relationship between sex and height. Also, since males have been designated by 0 and females by 1, the negative correlation means that females are generally shorter than males: Higher values for sex are associated with lower values for height. Given the teeny-tiny p-value, you can probably rest assured that males tend to be taller than females.

Shown in Figure 40.1 is the (weird looking) scatterplot with the A, B, C, and D regions. All the males' heights are bunched up at Sex = 0 and all the females'

Figure 40.1

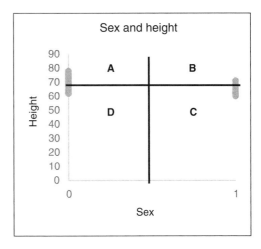

heights are bunched up at Sex = 1. You can see that the majority of males are above the overall average height and are in region A, and the majority of females are below the overall average height and are in region C. This makes a negative correlation.

41

A Simple Regression Prediction Model

Statistical Scenario—Height and Weight Regression

How can we come up with a way to actually make estimates and predictions?

Eyeballing our height and weight scatterplot from before, let's add a line that fits snugly inside the scatterplot data as shown in Figure 41.1. We'll call it a prediction line that predicts peoples' weights based on their heights.

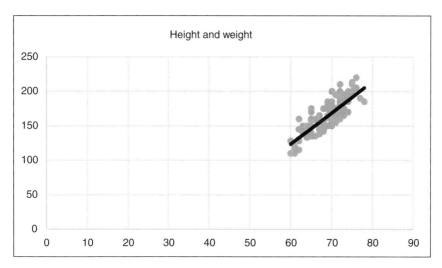

Figure 41.1

For any value of height (X) we can predict the value of weight (Y) using this line. For the average height of 69, the prediction is the average weight of 164 lbs. For 65 inches tall the prediction looks to be about 145 lbs. For 75 inches the

Illuminating Statistical Analysis Using Scenarios and Simulations, First Edition.
Jeffrey E Kottemann.
© 2017 John Wiley & Sons, Inc. Published 2017 by John Wiley & Sons, Inc.

prediction looks to be about 190 lbs. For less than 60 inches we probably should not hazard a guess, nor for more than 80 inches: Extrapolating outside the value ranges of our data is usually unwise.

Our eyeballed prediction line was placed visually to have the "snuggest fit" to the data. In order to develop a formulaic approach, we need to formally define what "snuggest fit" means. For the formulaic approach we'll be using, the snuggest fitting line is defined as the line with the *least* sum total of *squared* differences between the actual Y values and the Y values predicted by the line. (As usual, we square the differences to make them all positive.) In math shorthand, the method does the following[1]

$$\text{Minimize} \sum (\text{actual } Y \text{ value} - \text{predicted } Y \text{ value})^2$$

The method is appropriately called the least-squares method. Next, let's determine the least-squares line for our height and weight data step by step.

Recall from basic algebra that the formula for a line is

$$Y = a + bX$$

The line crosses the Y-axis at the Y-intercept, a, and the slope of the line is b.

Also recall from basic algebra that the slope is "rise over run." The rise/run formula that determines the slope, b, of the least-squares line is s_{xy} / s_x^2 which is the covariance of X and Y over the variance of X.

$$b = \frac{s_{xy}}{s_x^2}$$

When we plug in our height & weight sample covariance of 72.58 for s_{xy} and the height sample variance of 16.12 for s_x^2 we get 4.5 for b. That is the slope. An additional inch in height equates to an additional 4.5 pounds in weight on average.[2]

Next, we need to determine the Y-intercept, a. Least-squares lines always go through the intersection point of the two sample means, so we know the line goes through the coordinate (69, 164). And, we now know the slope is 4.5. So, with basic algebra we can solve for a via

$$a = \bar{y} - b\bar{x}$$

---

1 Predicted Y is often symbolized by $\hat{Y}$ ("Y hat"), but I'll continue to spell it out. The unsquared differences $(Y - \hat{Y})$ are called residuals. I don't use the term here, but it is a commonly used statistical term.

2 Given the formula, note that (1) the slope approaches zero when the covariance approaches zero, (2) the slope approaches zero when the variance of X approaches infinity, (3) the slope decreases when the variance of X increases, and (4) the slope is undefined when the variance of X equals zero.

When we plug in 164 for $\bar{y}$, 4.5 for b, and 69 for $\bar{x}$, we get -146.5 for a. So our formulaically derived least-squares prediction line for weight is

$$\text{Weight} = -146.5 + 4.5 \times \text{height}$$

Using the regression formula for a person who is 65 inches tall, we get a predicted weight of $-146.5 + 4.5 \times 65 = 146$ lbs, and for 75 inches we get $-146.5 + 4.5 \times 75 = 191$ lbs. Our earlier eyeballed estimates were 145 and 190.

> Is the slope b of 4.5 statistically significant? That is, significantly different from zero?

Let's use simulation to look at the distribution of values of b that can occur by chance. To mirror our real sample data, the simulation uses sample sizes of 120 as well as the sample means and variances of our height and weight data. Figure 41.2 shows the simulation results histogram for the slope, b, with a characteristic normal bell shape.

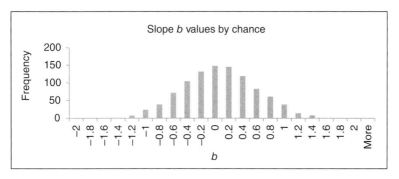

Figure 41.2

The Figure 41.2 histogram shows that it is very unlikely to get a value for b by chance that is outside the interval -1.4 to $+1.4$. The b value of 4.5 that we get with the real data is extremely unlikely to have arisen by chance alone. So, the p-value for our actual b must be extremely small. Based on this, we will reject the null hypothesis that b for the population equals zero.

The formulaic approach uses the <u>standard error of a regression coefficient</u> to determine how many standard errors b is away from zero:

$$\#\text{SEs} = b/\text{standard error of } b$$

The t-distribution is used to determine the p-value. Statistical analysis software gives a p-value of about 10^{-29} for b. The p-value for the intercept, a, is about 10^{-10}, although we often don't care about the value of the intercept or its statistical significance. In the current case, the intercept of -146.5 pounds is logically meaningless, but it is needed to fully define the line algebraically. There are cases, as we'll see in the next chapter, where the intercept is meaningful.

The method we are using is called simple least-squares linear regression and the prediction line is called the regression line or regression equation or regression model. The method is named with the word "simple" in the sense that it only has a single X variable.

Statistical analysis software will do all the least-squares regression computations and provides the sample correlation, the intercept and slope estimates along with their standard errors and p-values. The 95% confidence intervals for the intercept and slope estimates may also be provided.

Additional information typically provided by statistical analysis software will be highlighted as we cover the corresponding concepts.

42

Using Binomials Too

Statistical Scenario—Sex & Weight Regression

What does a simple linear regression model to predict weight using the binomial variable sex look like?

Let's look at the results of a simple regression model with the scaled variable weight as Y and the binomial variable sex as X. First, look at the (strange looking) scatterplot in Figure 42.1. All the male weights are bunched at $x = 0$ and all the female weights are bunched at $x = 1$. The regression line will go through the average weight for males and the average weight for females because that is the least-squares fit.[1]

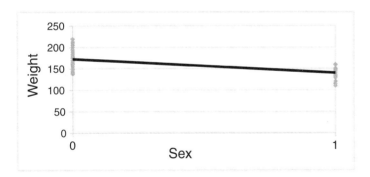

Figure 42.1

How can we interpret the slope of this regression line? Rise/run, where the rise is the difference between the sample mean weight for males and the sample mean weight for females, and the run is 1. The slope, then, is simply the

---

1 Binomials in regression models are often called "dummy variables" or "binary variables" or "indicator variables", among other names.

Illuminating Statistical Analysis Using Scenarios and Simulations, First Edition.
Jeffrey E Kottemann.
© 2017 John Wiley & Sons, Inc. Published 2017 by John Wiley & Sons, Inc.

difference in the sample means. That is what b will be. And the Y-intercept, a, is simply the average weight for males. With the data we get the following regression equation.

$$\text{Weight} = 172.78 - 33.71 \times \text{sex}$$

The average weight for the males in our sample is 172.78 pounds, and the average weight for the females in our sample is 33.71 pounds less than that. The sample correlation is -0.64; it is negative because higher values for sex are associated with lower values for weight. The p-value for r and for b are approximately 10^{-15}.

For a simple regression with a single binomial X variable:
The value of a, the Y-intercept, is the sample mean for the group with $x = 0$.
The slope, b, is the difference between the sample means for the two groups.

Since there is only one X variable in simple regression, r and b will have the same p-value. In the previous chapter, both had a p-value of 10^{-29}. In this chapter, both have a p-value of 10^{-15}. Also, the correlation between X and Y and the correlation between Y and predicted Y will have the same absolute value.

Getting More Sophisticated #1

(This can be skipped without loss of continuity.)

We will get the same p-value when we analyze the difference between two groups' sample means using: (1) the mean difference t-test of Chapter 27; (2) the one-way ANOVA method of Chapter 34; and (3) simple regression with a binomial X variable as in this chapter. After all, they are each evaluating the difference between sample means in one way or another. Table 42.1 shows the results of assessing the difference in mean weight for a random subset of 30 males and 30 females using each of the three methods.

Table 42.1 Showing equivalent results.

t-test	ANOVA	Regression	
$t = 5.36621106$	$F = 28.79622113$	$F = 28.7962211$	$t = 5.36621106$
p-value $= 0.0000015$	p-value $= 0.0000015$	p-value $= 0.0000015$	p-value $= 0.0000015$

Not to get too far down in the weeds, but notice that t squared equals F. That is because in general $F(1, df) = t^2(df)$.

Getting More Sophisticated #2

(This can be skipped without loss of continuity.)

Multinomials can also be incorporated into a regression model by using multiple binomial X variables. For example, let's say we want to model the opinion and political affiliation situation we saw in Chapter 34 on one-way ANOVA. Our Y variable will be the survey respondents' 1–7 scaled opinions. We will have two X variables: a binomial for Democrats, x_D, and another binomial for Republicans, x_R. We don't need a binomial variable for Independents because they are identified by having a value of zero for both x_D and x_R. Independents will therefore serve as the "reference" group. The regression model is a multiple regression in three-dimensional space (see the next chapter) rather than a simple regression in two-dimensional space:

$$Y = a + b_D x_D + b_R x_R$$

Table 42.2 gives example data points in this three-dimensional space.

Table 42.2 Three data points in three dimensions.

Y	x_D	x_R	
2	1	0	A Democrat with an opinion of 2
6	0	1	A Republican with an opinion of 6
4	0	0	An Independent with an opinion of 4

The regression's Y-intercept, a, will be the Independents' sample mean opinion. The slope b_D will be the difference between the Democrats' and Independents' sample means, and the slope b_R will be the difference between the Republicans' and Independents' sample means. The results of analyzing this regression model will mirror the ANOVA results we saw in Chapter 34.

More sophisticated regression model formulations allow a wide assortment of multidimensional analyses such as two-way ANOVA (Chapter 35). And, more sophisticated coding methods (rather than using the simple binomial variable values 0 or 1) can be used to support multifaceted group comparisons and contrasts. Do online searches for "multiple regression ANOVA table" and for "multiple regression ANOVA coding."

43

A Multiple Regression Prediction Model

Statistical Scenario—Height, Sex, and Weight Regression
How can we come up with a way to make use of both height and sex together to make (possibly) better predictions of weight?

Quick answer: Use <u>multiple linear regression</u>.

It is tricky to draw this situation, so I'll ask you to visualize it. Imagine a swarm of points shaped like a flying saucer, as in Figure 43.1.

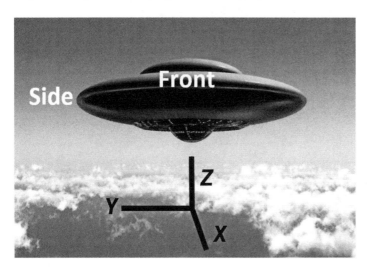

Figure 43.1

Now imagine a flat sheet of cardboard (or titanium) going straight through the middle separating the top of the flying saucer-shaped swarm of points from

Illuminating Statistical Analysis Using Scenarios and Simulations, First Edition.
Jeffrey E Kottemann.
© 2017 John Wiley & Sons, Inc. Published 2017 by John Wiley & Sons, Inc.

the bottom. That sheet snugly fits the points. The flying saucer and sheet can tilt along the X-axis by shifting the front up or down. The flying saucer and sheet can also tilt along the Y-axis by shifting the side up or down. With X and Y together, the flying saucer and sheet can have a combined tilt—up in front and down on the side; down in front and up on the side; and so on. The flying saucer and sheet can also levitate up and down along the Z-axis.

The sheet is a two-dimensional geometric plane snugly fitting into a three-dimensional swarm of points. It is the regression model. If we want to increase the number of dimensions to four or more (X, Y, Z, what?) we run out of letters. We need new notation: As shown in the new flying saucer Figure 43.2, we now use X_1, X_2, and so on and Y. Also, as shown in the below formula structure, b_0 is used to signify the Y-intercept and b_1, b_2, and so on are used to signify the slopes. The formula for a two-dimensional plane is a direct extension of the formula for a one-dimensional line.

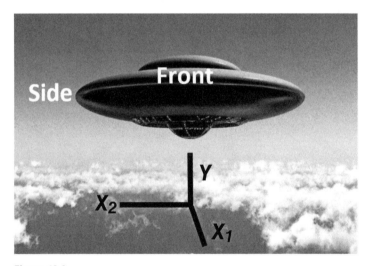

Figure 43.2

$$Y = b_0 + b_1 x_1 + b_2 x_2$$

Defines a geometric plane with Y-intercept b_0 and slopes b_1 and b_2

The tilts of the plane are governed by the two slopes b_1 and b_2 and the levitation up and down is governed by the Y-intercept b_0.

Table 43.1 shows least-squares multiple regression results for our data obtained via statistical analysis software.

Table 43.1 Multiple regression results.

Coefficient	B	Standard error	#SEs (*t*)	*p*-value	Lower 95%	Upper 95%
b_0 Intercept	−100.24	27.84	−3.60	<0.001	−155.38	−45.10
b_1 Height	3.87	0.39	9.82	<0.001	3.09	4.64
b_2 Sex	−9.20	3.60	−2.55	0.012	−16.33	−2.07
$R = 0.82$; ANOVA $F = 122.33$, *p*-value <0.001						

Looking at the Table's column labels, notice that we have seen all of these types of things before. The R shown at the bottom of the table is somewhat new: This is the correlation between the actual Y values and the predicted Y values. It indicates how snugly the regression model fits the data. R is always positive and ranges from zero to one. (With simple regression, R will equal the absolute value of r.)

The statistical significance (*p*-value) for the entire regression model, also shown at the bottom of the table, was determined via ANOVA with the null hypothesis that *all* the *slope* coefficients (b_1 and b_2) equal zero. In this case the *p*-value is less than our alpha-level of 0.05, so the overall regression model is statistically significant.

The sign of b_1 and of b_2 tell us whether the slope (tilts) of each X relative to Y is positive or negative (up or down) within the model.

The B column gives us the below prediction formula, with R of 0.82.

$$\text{Weight} = -100.24 + 3.87\text{height} - 9.20\text{sex}$$

The three separate null hypotheses for testing each of the three separate coefficients (b_0, b_1, b_2) are that the coefficient equals zero. The *t*-distribution is used for testing. As you can see in the table, each of the three coefficients has a value for #SEs (*t*) that yields *p*-values less than our alpha-level of 0.05, so each of the coefficients is statistically significant. All three null hypotheses are rejected. Also note that, since each of the three *p*-values is less than 0.05, each of the three 95% confidence intervals does *not* include zero.

Getting More Sophisticated

(This can be skipped without loss of continuity.)

The $-9.20 \times$ sex term implies that the actual Y-intercept for females (Sex = 1) is 9.20 lower than that for males (Sex = 0) when height is accounted for.

Given this, the intercept for males is b_0 and for females is $b_0 + b_2$ where b_2 is -9.20. So, we could construct separate prediction formulas for males and females as shown below.

$$\text{Male weight} = -100.24 + 3.87 \,\text{height}$$
$$\text{Female weight} = -109.44 + 3.87 \,\text{height}$$

While the model we are analyzing (Weight $= b_0 + b_1\text{height} + b_2\text{sex}$) accommodates different Y-intercepts for males and females, it assumes equal height slopes for males and females (an inch in height is 3.87 pounds in weight on average for both males and females). We could go further and add a term to accommodate different height slopes for males and females as follows.

$$\text{Weight} = b_0 + b_1\text{height} + b_2\text{sex} + b_3\text{height} \times \text{sex}$$

The term $b_3\text{height} \times sex$ is the <u>interaction</u> between height and sex. The height slope for males is b_1 and for females is $b_1 + b_3$. (With the present data, this interaction term is not statistically significant, with a p-value of 0.73).

Note: Our data are for adults. If instead the data were for children, then we may want to include age.

$$\text{Weight} = b_0 + b_1\text{height} + b_2\text{sex} + b_3\text{height} \times \text{sex} + b_4\text{age}$$

44

Loose End I (Collinearity)

Statistical Scenario—Loose End #1

$$\text{Weight} = -100.24 + 3.87\,\text{height} - 9.20\,\text{sex}; \quad R = 0.82$$

In Chapter 41 our simple regression model to predict weight using only height has R of 0.81, and in Chapter 42 our simple regression model to predict weight using only sex has R of 0.64.

In Chapter 43, our multiple regression using both height and sex has R of 0.82.

Why such a small improvement from 0.81 to 0.82 when we added sex to our height model?

Before we pursue this specific question, let's look at an extreme hypothetical example.

Consider Table 44.1. It tells us the number of bedrooms and bathrooms of new houses being built in the community along with the list price for each. There are many houses, but they are all one of the three types and they are all priced very close to their respective list prices.

Table 44.1 House features and prices.

X_1 (Number of bedrooms)	X_2 (Number of bathrooms)	Y (House list price)
2	1	100,000
3	2	150,000
4	3	200,000

Illuminating Statistical Analysis Using Scenarios and Simulations, First Edition.
Jeffrey E Kottemann.
© 2017 John Wiley & Sons, Inc. Published 2017 by John Wiley & Sons, Inc.

Let's see if $Y = b_0 + b_1 x_1$

and $Y = b_0 + b_2 x_2$

each predict just as well as $Y = b_0 + b_1 x_1 + b_2 x_2$

For the list prices, number of bedrooms and number of bathrooms are each linearly correlated with house price, and so either one can be used to predict house price alone: either $(0 + \text{bedrooms} \times 50000)$ or $(50000 + \text{bathrooms} \times 50000)$ predicts price.

Notice also that number of bedrooms and number of bathrooms are linearly correlated with each other: (Bedrooms $- 1 =$ bathrooms). Therefore, they are redundant predictors of house price. In this example—which is horrendously extreme—the two X variables are totally redundant. This is extreme collinearity, which means that pairs, *co*, of X variables can be defined as *linear* equations of each other.

If we used statistical software to analyze the actual house data, we would get a prediction formula something like this

$$Y = 0 + 50000 x_1 + 0 x_2$$

with the slope for x_2 of zero and therefore statistically nonsignificant. While this regression model works fine for prediction, the nonsignificance for the slope of x_2 is misleading: It is nonsense to interpret the regression model to mean that bedrooms are worth $50,000 each and that bathrooms are worth nothing. In fact, based on the correlations we know that both X variables are good predictors of Y, it is just that they are redundant predictors. In reality, collinearity is almost never this extreme, but it is fairly common in milder forms.

With our height, sex, and weight data, height and sex are somewhat redundant predictors of weight: Heavier people tend to be taller *and* male, and lighter people tend to be shorter *and* female. Table 44.2 shows the correlations of the three variables we are using.

Table 44.2 Correlations.

r	Height	Sex	Weight
Height	1		
Sex	−0.67	1	
Weight	0.81	−0.64	1

Looking at the weight row, we see the 0.81 correlation between height & weight, and the −0.64 correlation between sex and weight. In the sex row, we see the −0.67 correlation of height and sex, which are the two X variables in our regression model; they are correlated too. That is why adding sex to the weight $= b_0 + b_1$ height model does so little good.

Reproduced in Table 44.3 are our earlier multiple regression results.

Table 44.3 Multiple regression results.

Coefficient	B	Standard error	#SEs (t)	p-value	Lower 95%	Upper 95%
b_0 Intercept	−100.24	27.84	−3.60	<0.001	−155.38	−45.10
b_1 Height	3.87	0.39	9.82	<0.001	3.09	4.64
b_2 Sex	−9.20	3.60	−2.55	0.012	−16.33	−2.07
$R = 0.82$; ANOVA $F = 122.33$, p-value <0.001						

The simple regression involving sex and weight (Chapter 42) has −33.71 × sex, but this multiple regression has −9.20 × sex. The p-value for sex in the simple regression is 10^{-15} but the p-value in the multiple regression is 0.012. This is what happens when collinearity is present. In fact, as shown below, if the collinearity in the data is a bit more extreme, the p-value for sex will cease to be statistically significant.

To illustrate, I altered the height data slightly to strengthen the correlation between height and sex. (This was done simply by subtracting a small *random* amount from each female's height, which strengthens the sex to height negative correlation without affecting the other correlations.) As shown in Table 44.4, only the sex to height correlation has changed, from −0.67 to −0.75.

Table 44.4 New correlations.

r	Alt. Height	Sex	Weight
Alt. Height	1		
Sex	−0.75	1	
Weight	0.81	−0.64	1

Table 44.5 New multiple regression results.

Coefficient	B	Standard error	#SEs (t)	p-value	Lower 95%	Upper 95%
b_0 Intercept	−88.42	27.71	−3.19	0.002	−143.30	−33.55
b_1 Alt. Height	3.70	0.39	9.44	<0.001	2.92	4.48
b_2 Sex	−3.57	4.12	−0.87	0.389	−11.73	4.60
$R = 0.82$; p-value <0.001						

Now, look at the new regression results in Table 44.5. Now the *p*-value for sex is 0.389 and is no longer statistically significant. This could lead us to erroneously surmise that sex is not related to weight (which is the null hypothesis), when in fact it is. Looking only at these multiple regression results, we would not reject the null hypothesis even though it is actually false. This would constitute a type II error.

> Collinearity increases the chances of type II error for redundant predictors.

Collinearity will also cause the values of the *b* coefficients in a multiple regression model to be misleading. Recall the extreme case from above, where the regression model for house prices seemed to suggest that bedrooms were worth $50,000 each and that bathrooms were worth nothing at all. This is why you should always look at the correlations of all the *X* variable pairings in addition to looking at the multiple regression results. That helps you properly interpret the results. It is not foolproof though. More subtle multicollinearity may exist.

Multiple *X* variables, rather than simply pairs, may also be linearly related. Professional grade statistical analysis software typically has measures to help assess multicollinearity, with the most common being variance inflation factor (VIF) and tolerance. Do an online search with those terms for more information.

45

Loose End II (Squaring R)

R and r are scaled in terms of standard deviations—notice that the correlation formula in Chapter 37 has standard deviations as divisors. To rescale R in terms of variance, we simply square it. For R of 0.82 we get R^2 of about 0.68. R^2 is the proportion of the variance in our sample weights that is "explained by" the regression model. We would say "our regression model explains 68% of the variance in sample weights."

R^2 is the proportion of variance in the Y variable that is explained by the regression model. The official name for R^2 is the coefficient of determination, but everyone just calls it R-squared.

R and R-squared indicate how closely packed the data points are to the regression line or plane. To get a visual feel for this, and to compare R and R-squared, look at the scatterplots showing simple regression lines in Figures 45.1 and 45.2.

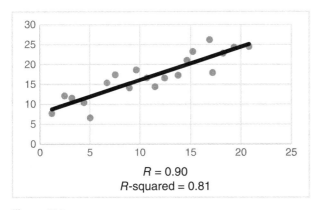

$R = 0.90$
R-squared $= 0.81$

Figure 45.1

Illuminating Statistical Analysis Using Scenarios and Simulations, First Edition.
Jeffrey E Kottemann.
© 2017 John Wiley & Sons, Inc. Published 2017 by John Wiley & Sons, Inc.

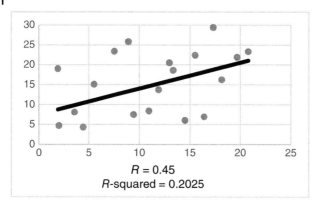

$$R = 0.45$$
$$R\text{-squared} = 0.2025$$

Figure 45.2

46

Loose End III (Adjusting *R*-Squared)

We need one more tweak.

R^2 can be a bit misleading when we have multiple X variables, because each additional X variable we add to a model increases the possibility that *chance correlations* with Y will occur. Adjusted *R*-squared adjusts for that possibility.

$$\text{Adjusted } R^2 = 1 - \frac{(1 - R^2)(n - 1)}{n - \#Xs - 1}$$

Since the number of X variables, $\#Xs$, is subtracted in the denominator, adjusted R^2 decreases as $\#Xs$ increases. In certain cases (small R^2 or small n or large $\#Xs$) adjusted R^2 can have a slightly negative value that should be interpreted as zero.

Let's check using random numbers for data

1) When we perform simple regressions with Y and one X variable whose data values are all random numbers, R^2 averages about 0. This is what you would expect because random numbers should not be able to predict anything, much less other random numbers. Adjusted R^2 likewise averages about 0.
2) But, when we perform multiple regressions with Y and eight separate X variables, $X_1, X_2, \ldots$, and X_8, whose data values are all random numbers, R^2 averages about 0.05. Why? Because the more X variables you use, the more likely it is that some of the X variables will just happen to correlate with Y by accident. Adjusted R^2 adjusts down from R^2 to compensate. For this situation with eight X variables of random numbers, adjusted R^2 averages about 0.

Illuminating Statistical Analysis Using Scenarios and Simulations, First Edition.
Jeffrey E Kottemann.
© 2017 John Wiley & Sons, Inc. Published 2017 by John Wiley & Sons, Inc.

With our two X variable model

Weight $= -100.24 + 3.87$ height -9.20 sex; $R = 0.82$; $R^2 = 0.68$; Adj $R^2 = 0.67$

Adjusted R^2 is just a little less than plain R^2. Using three decimal places, plain R^2 is 0.676 and adjusted R^2 is 0.671.

You can use adjusted R^2 to judge the goodness of fit of a regression model.

47

Reality Strikes

You may recall that the data we have been using has been trimmed of "extreme" cases so that scatterplot patterns and correlation and regression results would jump out at us. Table 47.1 shows the multiple regression results again.

Table 47.1 Multiple regression results from before.

Coefficient	B	Standard error	#SEs (t)	p-value	Lower 95%	Upper 95%
b_0 Intercept	−100.24	27.84	−3.60	<0.001	−155.38	−45.10
b_1 Height	3.87	0.39	9.82	<0.001	3.09	4.64
b_2 Sex	−9.20	3.60	−2.55	0.012	−16.33	−2.07
$R = 0.822$; $R^2 = 0.676$; Adj $R^2 = 0.671$; p-value <0.001; $n = 120$						

$$\text{Weight} = -100.24 + 3.87 \times \text{height} - 9.20 \times \text{sex}; \quad \text{Adj } R^2 = 0.671$$

I don't want to leave you with a false sense of reality, so Table 47.2 shows the results using the full sample of 186 people.

Table 47.2 Multiple regression results using all the data.

Coefficient	B	Standard error	#SEs (t)	p-value	Lower 95%	Upper 95%
b_0 Intercept	−29.74	47.54	−0.63	0.532	−123.54	64.06
b_1 Height	3.00	0.67	4.46	<0.001	1.67	4.33
b_2 Sex	−24.30	6.03	−4.03	<0.001	−36.19	−12.41
$R = 0.584$; $R^2 = 0.341$; Adj $R^2 = 0.334$; p-value <0.001; $n = 186$						

Illuminating Statistical Analysis Using Scenarios and Simulations, First Edition.
Jeffrey E Kottemann.
© 2017 John Wiley & Sons, Inc. Published 2017 by John Wiley & Sons, Inc.

$$\text{Weight} = -29.74 + 3.00 \times \text{height} - 24.30 \times \text{sex}; \quad \text{Adj } R^2 = 0.334$$

In reality, height and sex in our overall sample don't do all that great a job predicting weight. They only explain about 33% of the variance of peoples' weights in this analysis. Notice that the standard errors have nearly doubled. The scatterplot for height and weight of the 120 people that we have used up until now is reproduced in Figure 47.1 and the scatterplot for the full sample of 186 people is shown in Figure 47.2. Look at the differences. With the full sample, could the 6'5" 350 lbs person be unduly affecting the regression results? What about the 5'4" 250 lbs person? (Note that 5'4" is not an extreme value for height, and 250 lbs is not an extreme value for weight, but the *combination* of the two is extreme.)

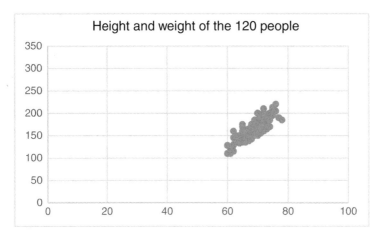

Figure 47.1

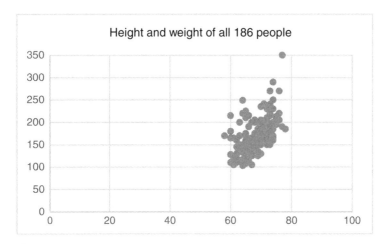

Figure 47.2

Among the most insidious problems that arise when analyzing scaled data are due to extreme values, called <u>outliers</u>. With the least-squares method we are using, recall that the snuggest fitting line is defined as the line with the least sum total of *squared* deviations between the predicted Y values and the actual Y values. Since it squares the deviations, and since outliers represent potentially large deviations, the squared deviations of outliers can be overly influential: The least-squares method will shift the line toward the outlier, perhaps dramatically.

Take a look at the scatterplots with the regression lines shown in Figures 47.3a and 47.3b. The only difference is that an extreme outlier exists in 47.3b. The R^2 in Figure 47.3a is about 0.68. The R^2 in Figure 47.3b with the outlier is about 0.00.

Figure 47.3

(a)

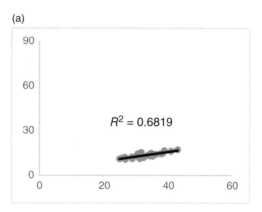

(b)

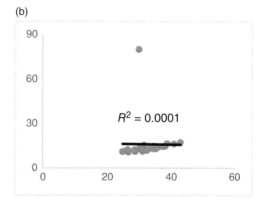

Outliers can make it seem that relationships don't exist when they really do.

Next, take a look at the scatterplots in Figures 47.4a and 47.4b. Again, the only difference is an extreme outlier in 47.4b. The R^2 in Figure 47.4a is only about 0.02. The R^2 in Figure 47.4b with the outlier is about 0.97.

(a)   **Figure 47.4**

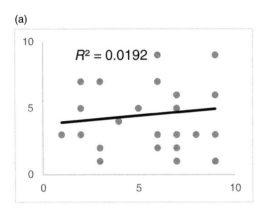

(b)

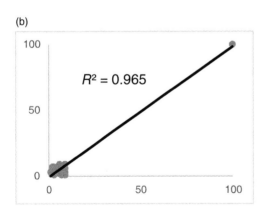

Outliers can make it seem that relationships do exist when they really don't.

Outliers can seriously undermine the statistical analysis methods we have looked at for scaled data. Sometimes outliers are not a problem, or not the only problem. What if 90% of the sample data have relatively low values while the other 10% are spread out over higher and higher values? Are they outliers or just part of a peculiar distribution? Regardless, you can't selectively omit 10% of your data and still call it unbiased evidence. Part V explores issues like these.

This marks the end of Part IV. At this point you can look at the review and additional concepts in Part VI Chapter 59, or proceed directly to Part V.

Part V
Dealing with Unruly Scaled Data

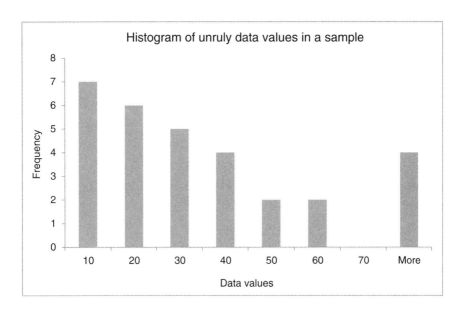

Illuminating Statistical Analysis Using Scenarios and Simulations, First Edition.
Jeffrey E Kottemann.
© 2017 John Wiley & Sons, Inc. Published 2017 by John Wiley & Sons, Inc.

48

Obstacles and Maneuvers

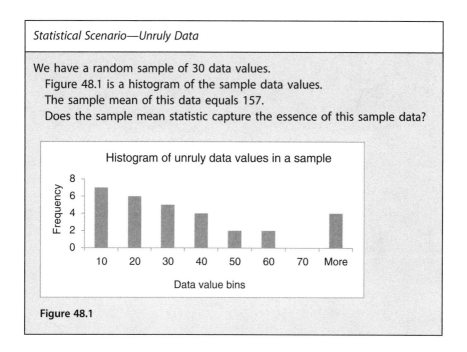

Statistical Scenario—Unruly Data

We have a random sample of 30 data values.
Figure 48.1 is a histogram of the sample data values.
The sample mean of this data equals 157.
Does the sample mean statistic capture the essence of this sample data?

Histogram of unruly data values in a sample

Figure 48.1

The formulaic methods for scaled data in Parts II, III, and IV assume that sample data itself is normally distributed, or fairly close to it. And while this assumption is often warranted, sometimes it is not. The *statistical scenario—unruly data* is a case in point. That data harbors two nonnormal traits that are common obstacles to using the statistical analysis methods we've seen so far for scaled data.

Illuminating Statistical Analysis Using Scenarios and Simulations, First Edition.
Jeffrey E Kottemann.
© 2017 John Wiley & Sons, Inc. Published 2017 by John Wiley & Sons, Inc.

The first obstacle, as you know, are outliers. Figure 48.2a shows the impact of one extreme outlier in a sample of size 30 (the outlier value of 1000 shows up as a little nub in the "More" slot). Figure 48.2b shows the data with the outlier removed. Notice the large impact the outlier has on the sample mean and the even larger impact it has on the sample variance that in turn will wreak havoc on various other sample statistics, confidence intervals, and significance tests.

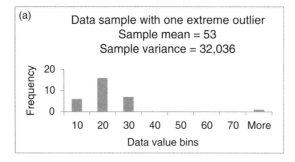

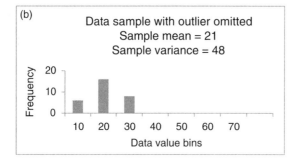

Figure 48.2

Statistical analysis software typically has features to help identify outliers; at the very least it will highlight the minimum and maximum values in the sample data. When researchers can justify omitting or adjusting outlier(s), they sometimes do so, making a note of what they did for all to see. If not, there are other strategies, as we'll see.

Another common obstacle is skewed data. As shown in the Figure 48.3 data histogram, skewed distributions are nonsymmetrical distributions. Statistical analysis software typically has features to help identify skew. There is even a skewness statistic that equals zero for nonskewed (symmetric) distributions. The skewness statistic also helps detect the presence of outliers, since outliers may skew only to one side or the other. When sample data is distributed fairly normally, the skewness statistic will be between −1 (left skew) and +1 (right skew). By this criterion, the data of Figure 48.3 is not "fairly normal".[1]

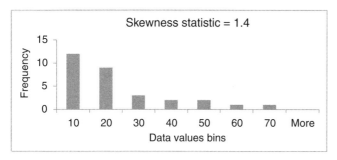

Figure 48.3

There are at least two general strategies to remedy skew. The first involves rescaling the data using, for example, logarithms. We'll save logarithms for later. The second general strategy can be used to address both outliers and skew: transform the data into a different type of data. (Appendix A overviews the general types of data.)

This second general strategy involves transforming the scaled data to nominal or ordinal. We can reformulate all the scaled data into categories, redefining the variables to be binomial such as "income at or below the poverty line" and "income above the poverty line" or multinomial such as "income below 24,999; income from 25,000 to 49,999; . . . ; and income over 199,999". We could then use the various methods suited for nominal variables. However, when converting to nominal we'll lose quite a bit of the information contained in the scaled data. Also, the choice of category cutoffs can introduce bias.

---

1 There is also a statistic called the kurtosis statistic that helps identify data distributions that are too bunched up or are too spread out. With a normal distribution, the kurtosis statistic equals 3 and the excess-kurtosis statistic equals 0. When data is fairly normal, the excess-kurtosis statistic will also be between −1 and +1. These limits for skew and kurtosis *are just rules of thumb*. There are also formal statistical tests for data normality, but those have limitations of their own (see, for example, the last section of Chapter 30). To add to the confusion, different statistical inference methods (*t*-test, *F*-test, etc.) have different sensitivities to violations of data distribution assumptions and therefore can have different rules of thumb. Also because of this, there is an assortment of statistical test alternatives that have been developed to address the violations.

A potentially better alternative involves transforming the scaled data into ordered rankings. As we'll see, this will help even if our data looks like Figure 48.4.

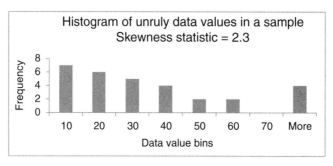

Figure 48.4

49

Ordered Ranking Maneuver

Let's look at some scaled numbers with their rank numbers shown below them—see Table 49.1.

Table 49.1 Scaled data values and their rank numbers.

Scaled number	1.0	1.2	1.8	2.1	2.1	2.2	2.3	3.2	4.0	5.9	29.5	49.2	1000.0
Rank number	1	2	3	4.5	4.5	6	7	8	9	10	11	12	13

The scaled numbers are put in sorted order, and then rank numbers are assigned in that order. Notice that in the case of ties, the local ranks are averaged and shared, as with the $(4 + 5)/2 = 4.5$ shared rank.

The scaled data value associated with the middle rank is called the <u>median</u>, which has the scaled number value of 2.3 in Table 49.1. The median is a good alternative statistic to the mean when data distributions are unruly because the median is unaffected by outliers and skew. The mean of the numbers in Table 49.1 is 85, pulled drastically up by skew and the extreme value of 1000. Further, if the value of 1000 were 10,000 instead, the mean would become 777. But the median would remain 2.3. And, importantly, the rank number 13 would remain 13.

Using rank numbers literally brings unruly scaled data distributions into line. Scaled data is transformed into ordinal rank data, and, as we'll see, *the rank numbers can be analyzed in place of the original scaled numbers.*

Conveniently, with ranks we can calculate certain "reference values." For example, the <u>sum of ranks</u> 1– through n will equal $n(n + 1)/2$. For $n = 100$, that is 5,050. And, when the 100 items in a sample are comprised of two identically ranked groups of size 50 each, they will each have a <u>rank sum</u> in the middle of that: $5050/2 = 2525$. If the two groups are different, one will have a rank sum less than 2525 by a certain amount and the other will have a rank sum greater than 2525 by that same amount. Next, we'll see how all this comes in handy.

Illuminating Statistical Analysis Using Scenarios and Simulations, First Edition.
Jeffrey E Kottemann.
© 2017 John Wiley & Sons, Inc. Published 2017 by John Wiley & Sons, Inc.

50

What Rank Sums Happen Just by Chance?

Statistical Scenario—Rank Sums Distribution
What will the distribution of sample rank sums look like? How can we determine confidence intervals and p-values?

Let's experiment with using rank numbers for statistical inference. We'll simulate a generalized version of the following scenario: We have annual income data for 100 random people, a group of 50 males and a group of 50 females. The data distribution is very unruly, with skew and outliers too, so we sort and rank them 1–100 by annual income. We'll presume that there is no population difference between males and females in annual income rankings; this is the null hypothesis.

As we saw in the previous chapter, summing all the ranks of 1–100 gives 5050. If the males and females in the sample have identical rankings, then they will both have a rank sum of half that: 2525. If they are different, one will have a rank sum less than 2525 by a certain amount and the other will have a rank sum greater than 2525 by that same amount. So, we only need to look at one of the rank sums. We can sum the actual ranks of males, for example, and see how far away it is from 2525. How far away is far enough away to reject the null hypothesis?

The simulation we'll perform for this scenario draws a sample of 50 random male rank numbers from the set of possible rank values 1–100, and then sums the 50 randomly drawn ranks. This gives us a male sample rank sum determined by chance alone. We'll do this 10,000 times.

Figure 50.1 is the simulation histogram showing the 10,000 sample rank sums. It shows what to expect when the null hypothesis is true.

The reason for the normality of the rank sum distribution is analogous to what we have seen before: For any given sample, it is very unlikely to randomly select most or all low ranks and it is unlikely to randomly select most or all high

Illuminating Statistical Analysis Using Scenarios and Simulations, First Edition.
Jeffrey E Kottemann.

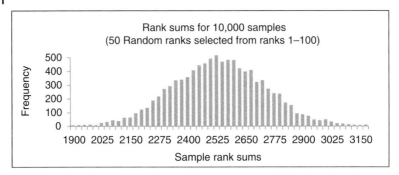

Figure 50.1

ranks, and so the sum will most likely be toward the middle of the possible range. With the Figure 50.1 histogram, the middle is where it should be: 2525.

The Figure 50.1 histogram shows that it is fairly unlikely to have a rank sum less than 2150 or more than 2900 if the null hypothesis is true. The corresponding interval of 2525 ± 375 appears to be the approximate 95% confidence interval. If the actual male rank sum is outside that interval we can reject the null hypothesis at the 95% confidence level. (If the male rank sum is outside the interval on one side, the females rank sum will be outside the interval on the other side.)

We can also derive a *p*-value based on how far out a rank sum is from the middle. For example, let's say the male rank sum is 2025. Using the Figure 50.1 histogram, about 80 of the 10,000 rank sums are less than or equal to 2025, which translates to a one-tailed *p*-value of 0.008 and a two-tailed *p*-value of 0.016.

This illustrates the general essence of using ranks for statistical inference. Next, let's look at a corresponding formulaic method.

51

Judging Rank Sum Differences

Let's look at a formulaic method that corresponds to what we just did (and is named after the originators, Mann, Whitney, and Wilcoxon). We'll be testing for a rank sum difference between 30 randomly selected males and 30 randomly selected females. The female sample contains an extreme outlier. We'll look at the statistical analysis results produced by statistical analysis software, and walk through what it did.

But first, look at Table 51.1 to see how the standard t-test is bamboozled by the unruly data: Notice the impact of the outlier on the female sample mean, variance, number of standard errors, and p-value. This is what happens when you have unruly scaled data and use methods from Parts II, III, and IV.

Table 51.1 An inappropriate t-test.

t-Test for mean difference—*Whoops!*		
	Males	**Females**
Mean	15.5	*119.3333*
Variance	77.5	*296083.3*
Observations	30	30
Hypothesized mean difference	0	
df	59	
t (#SEs)	*−1.04504*	
p-value, two-tail	*0.304637*	

Illuminating Statistical Analysis Using Scenarios and Simulations, First Edition.
Jeffrey E Kottemann.
© 2017 John Wiley & Sons, Inc. Published 2017 by John Wiley & Sons, Inc.

Now, look at Table 51.2 to see how the rank sum test takes the unruly data in stride. The statistical analysis software took the scaled data, constructed corresponding rank data, and then performed the Mann–Whitney–Wilcoxon rank sum test using the rank data.

Table 51.2

MWW rank sum test		
Male sample size	30	
Sum of male ranks	777.5	
Female sample size	30	
Sum of female ranks	1052.5	
Total sample size n	60	a
Male rank sum	777.5	b
Middle rank sum	915	c
Difference	−137.5	d
Standard error	67.64	e
z (#SEs) Test statistic	−2.03	f
p-Value (two-tail)	0.04	g

It is straightforward to walk through the logic. The total sample size is 60 (a). The males have an actual rank sum of 777.5 (b). If males and females were the same, they would both have rank sums equal to the middle rank sum of 915 (c). The difference between the male rank sum and the middle rank sum equals −137.5 (d). The standard error of a sample rank sum is 67.64 (e).[1] The actual difference of −137.5 is rescaled by dividing it by the standard error of 67.64 yielding −2.03 standard errors (f). Since rank sums are normally distributed and since standard error is determined solely by sample sizes, the z-distribution can be used to get the p-value. As we know, 2.03 on the z-distribution will give us a p-value of a little less than 0.05. The p-value determined by the software is 0.04 (g).

---

1 Standard Error is $\sqrt{\frac{1}{12}n_1 n_2 (n_1 + n_2 + 1)}$ when there are no ties.

For this example it is $\sqrt{\left(\frac{1}{12}\right) \times 30 \times 30 \times (30 + 30 + 1)} = 67.6387$.

Notice that only sample sizes are needed for the calculation of standard error. The formula is more involved when ties are accounted for.

52

Other Methods Using Ranks

When data is unruly, using ranks in place of the original scaled data is a good maneuver. And just as the Mann–Whitney–Wilcoxon method uses ranks in place of scaled data and provides an alternative to the t-test for mean differences, there are other methods that use ranks and serve as alternatives to other methods we've seen.[1]

- Methods to construct confidence intervals for the median (rather than the mean).
- The Kolmogorov–Smirnov test helps determine whether a sample is consistent with a specified distribution and serves as an alternative to the Chi-squared goodness of fit test (Chapters 28 and 30).
- The Kruskal–Wallis method is an extension of the Mann–Whitney–Wilcoxon rank sum method and offers an alternative to one-way ANOVA (Chapter 34).
- Spearman's rank correlation coefficient and Kendall's tau rank correlation coefficient are alternatives to Pearson's correlation coefficient (Chapter 37).

As for alternatives for least-squares regression, often the best bet is to transform the data to another numeric scale or to transform the least-squares criteria itself through use of robust regression methods. These are our next topics.

---

1 Methods that use ranks are often referred to as <u>nonparametric methods</u>.

Illuminating Statistical Analysis Using Scenarios and Simulations, First Edition.
Jeffrey E Kottemann.
© 2017 John Wiley & Sons, Inc. Published 2017 by John Wiley & Sons, Inc.

53

Transforming the Scale of Scaled Data

Heavily skewed distributions are actually somewhat common in some contexts. For example, income across households as well as gross domestic product (GDP) per capita across countries are not normally distributed. Such distributions are heavily skewed to the right (positively skewed). In the case of household income, for example, the income of a large majority of households falls between zero and, say, $75,000 and a minority of richer households are spread out all the way to who knows how much. Economists typically transform such values by taking their (natural) logarithms, which helps to normalize skewed data distributions. They can then use the transformed values in their linear regression models.

Logarithmic and exponential functions are inverses of each other; they undo each other. For example: $\log_{10}10 = 1$ and $10^1 = 10$; $\log_{10}100 = 2$ and $10^2 = 100$; $\log_{10}1000 = 3$ and $10^3 = 1000$; and so on. When taking logs, the larger an original value is, the more it is decreased relative to smaller original values. The same is true when we take square roots, which can be used when skew is less pronounced. Both logs and square roots are <u>nonlinear</u> transformations.

Figure 53.1a&b shows the "before" and "after" histograms of a log transformation.

There are different types of transformations that can be used in different situations. Do an online search for "data transformation to normal".

Illuminating Statistical Analysis Using Scenarios and Simulations, First Edition.
Jeffrey E Kottemann.

(a)

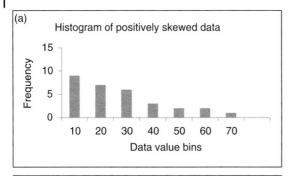

(b)

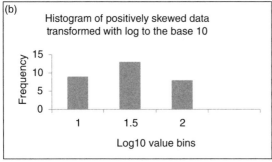

Figure 53.1

54

Brief on Robust Regression

Rather than transforming the data, we could transform certain aspects of the statistical analysis method. For example, what would happen if we transformed the "least-squares" criteria in least-squares linear regression into a "least absolute value" criteria instead?

Instead of this (Chapter 41)

$$\text{Minimize} \sum (\text{actual } Y \text{ value} - \text{predicted } Y \text{ value})^2$$

we have this

$$\text{Minimize} \sum |(\text{actual } Y \text{ value} - \text{predicted } Y \text{ value})|$$

The impact of outliers would certainly be reduced. Recall that squaring the deviations between the actual and predicted values for Y is what made outliers so extra troublesome for least-squares linear regression because least-squares regression fits the regression equation to minimize the squared deviations. If we used the absolute value of the deviations instead, the impact of outliers would be less severe.

This is the type of thing that is done by robust regression methods. Do an online search for "robust regression". A related topic is "generalized linear models".

Illuminating Statistical Analysis Using Scenarios and Simulations, First Edition.
Jeffrey E Kottemann.
© 2017 John Wiley & Sons, Inc. Published 2017 by John Wiley & Sons, Inc.

55

Brief on Simulation and Resampling

<u>Simulations</u> such as we have been doing can be used much as we have been using them. Simulations are particularly useful when the assumptions that must be met in order to use formulaic methods are not met. Theoreticians who study statistical phenomena and the detailed workings of statistical methods often use simulation; they typically use more powerful tools than spreadsheet software, such as the open-source statistical toolkit R.

<u>Resampling</u> is analogous to the simulations we have been doing, but instead of repeatedly sampling random numbers in accordance with a preselected distribution, we would repeatedly (re)sample randomly selected values from a real data sample itself. Let's see how such an approach might work. The following example illustrates a popular approach to resampling called <u>bootstrapping</u>.

Say that we have a very unruly random sample of 500 incomes. We want to determine a 95% confidence interval for the population mean. We could do the following 1000 times: (1) randomly select (resample) 500 incomes from the overall sample, allowing any income to be selected any number of times, and (2) calculate the mean and save it on a list. Then, from the list of 1000 resample means, use the 25th smallest resample mean and 25th largest resample mean as the boundaries for a 95% confidence interval. Alternatively, or in addition, the standard deviation of the 1000 resample means can be used as an estimate for standard error (see Appendix C).

Simulation and resampling methods can be used as alternatives or augmentations to many statistical methods and are supported (to varying degrees) by professional grade statistical analysis software. Do online searches for "statistical simulation" and for "statistical resampling".

> This marks the end of Part V. At this point you can look at the short review in Part VI Chapter 60, or proceed directly to Appendices E, F, and G.

Part VI
Review and Additional Concepts

The *statistical scenarios* in Part VI are structured as questions and answers. The Q&A reinforces previously covered concepts as well as introduces new concepts. There are some *statistical scenarios* that involve informal or formal statistical analysis, and others that involve interpreting the results of statistical analysis. We shall also look at some important practical issues, such as the difference between *statistical* significance and *practical* significance.

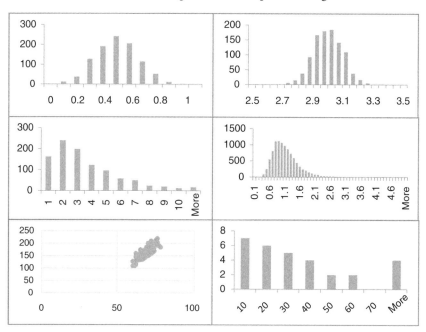

Illuminating Statistical Analysis Using Scenarios and Simulations, First Edition.
Jeffrey E Kottemann.
© 2017 John Wiley & Sons, Inc. Published 2017 by John Wiley & Sons, Inc.

56

For Part I

(A) Figure 56.1 is a histogram showing the results of 1000 repetitions of flipping a fair coin 100 times. The results are the number of times the 100 flips came up with various proportions of heads. Does this look approximately like a normal distribution? Eyeballing this histogram, what is the approximate 95% confidence interval for a fair coin? What is the approximate 99% confidence interval?

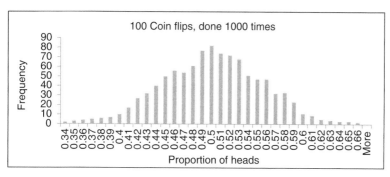

Figure 56.1

> *Answer:* Yes, it approximates a normal distribution. It looks like about 950 of the 1000 proportions (95%) are in the interval 0.4–0.6. About 50 (5%) of the results have proportions outside this interval, with about 25 (2.5%) on each side (in each tail). Also, although it is hard to eyeball, about 990 of the 1000 proportions (99%) are betwen about 0.37 and about 0.63 inclusive, with about 5 (0.5%) in each tail outside this interval. (Simulating 10,000 repetitions would be better.)

(B) Looking at Table 56.1, check whether your eyeballed 95% confidence interval matches the theoretical results one should obtain when flipping

Illuminating Statistical Analysis Using Scenarios and Simulations, First Edition.
Jeffrey E Kottemann.
© 2017 John Wiley & Sons, Inc. Published 2017 by John Wiley & Sons, Inc.

216 | 56 For Part I

fair coins 100 times. (Recall that the margin of error is the half-width of the 95% confidence interval and can be expressed as a percentage rather than a proportion.)

Table 56.1

Sample size	100	500	1000	1500
Margin of error (%)	±9.80	±4.38	±3.10	±2.53

Answer: Yes, it does. 50% ± 9.8% is about 40–60%, or 0.4–0.6 proportions.

(C) If you were going to survey 100 random people and your presumption was that the community is evenly split, what would be your 95% confidence interval?

Answer: Same as with the fair coin: proportions from 0.4–0.6.

(D) Calculate the 95% and 99% confidence intervals using the formula. (Recall that you replace the 1.96 standard errors with 2.58 standard errors to get the 99% confidence interval.)

$$\text{From } p - 1.96\sqrt{(p \times (1 - p))/n} \text{ to } p + 1.96\sqrt{(p \times (1 - p))/n}$$

Answer: Plug in $p = 0.5$ and $n = 100$ and do the arithmetic. Rounding to the 100ths place we get 0.40–0.60 and 0.37–0.63. (Effectively the same as our eyeballed intervals.)

(E) If your survey results were, say, 35% agree, would you retract your presumption of the community being evenly split? Use a 95% confidence level (0.05 alpha-level). What about when using a 99% confidence level (0.01 alpha-level)?

Answer: Yes, reject the null hypothesis of the community being evenly split. In fact, 0.35 is even outside the 99% confidence interval.

(F) Let's say another surveyor doing a different survey found that 53 people out of 100 surveyed agree with a new community policy. The local newspaper wrote "a recent survey found that a slim majority of community members agree with the new policy." Should the newspaper have published that?

Answer: No. 53% agree on a survey of 100 people is not sufficient evidence to reject the presumption of an evenly split community. The 95%

confidence interval for evenly split is 50±9.8%, which certainly contains 53%. Survey results are misinterpreted like this frequently. Another common misinterpretation is "the community leans slightly toward agree" which is also incorrect. The survey does not provide evidence for anything except something like "the survey does not indicate that the community leans one way or the other". (Not a very catchy headline.)

(G) Let's say another surveyor doing a different survey found that 530 people out of 1000 (53%) surveyed agree with a new community policy. The local newspaper wrote "a survey found that a majority of community members agree with the new policy." Should the newspaper have written that?

Answer: No. If you look at Table 56.1, you will see that the margin of error for a sample size of 1000 is ±3.1%. Even with a sample size of 1000, 53% agreeing is not sufficient evidence to reject the presumption of an evenly split community. (Either way you look at it: 50±3.1% contains 53% and 53±3.1% contains 50%)

(H) How about with a sample size of 1500?

Answer: Well, then, yes. Using Table 56.1, the 95% confidence interval for evenly split is 50±2.53%, and the survey's 53% agree is outside that interval. But the newspaper should note that there is uncertainty and give the margin of error for the survey. Also, the choice of confidence level is critical in all this. The 95% confidence level (alpha-level of 0.05) is an accepted convention in public opinion polling, and many other contexts, but the choice of 95% is just that, a choice. Also, recall Chapter 17 and the false discovery rate. Given the huge number of public opinion polls that are seemingly always being conducted, the false discovery rate phenomenon is always important to keep in mind.

(I) You are a particle physicist smashing protons into each other. You are looking for evidence of a new particle. You have been gathering evidence by smashing protons into each other for weeks and have the p-value down to 1 in 10,000 (0.0001). Do you announce that you have discovered a new particle?

Answer: No, keep smashing. By the conventions of particle physics, 1 in 3.5 million (5 standard errors) is the alpha-level to use. You need a p-value less than about 0.0000003.

(J) The community survey also yielded demographic information. Of the 1500 random people surveyed, 53% were woman. Can you reject the null hypothesis that the community is evenly split by sex at the 95% confidence level?

Answer: Yes (if you are sure it was a random sample), but $53 \pm 2.53\%$ women certainly does not suggest that there is an overwhelming sex imbalance in the community.

(K) Surveyors surveyed 1000 people as they entered and exited the community shopping mall. 70% of those surveyed said they agree with the new community policy. That is well outside the 99% confidence interval. Can it be said "the survey results suggest that a majority of community members agree with the new policy?"

Answer: No. It is not a random sample. It is biased toward the opinions of people who frequent the shopping mall: that is called <u>selection bias</u>. The results can not be generalized to all community members.

(L) This time the surveyors used a list of all the phone numbers in the community and selected random phone numbers to call. They made calls until they garnered 1000 peoples' survey responses. How is that?

Answer: Much better than surveying at the shopping mall. However, don't some "types of people" tend to answer their phone, and other "types of people" tend not to? That is a potential source of selection bias. It is often very tricky to get a truly random sample, and there are many issues involved in sampling. Do an online search for "statistical sampling."

(M) A survey of 1000 random people yielded a sample proportion agree of 0.30. Use the formula to construct the 95% confidence interval for the true population proportion.

$$\text{From } p - 1.96\sqrt{(p \times (1 - p))/n} \text{ to } p + 1.96\sqrt{(p \times (1 - p))/n}$$

Answer: Plug in 0.30 for p, 1000 for n. This gives you from 0.271596902 to 0.328403098. You will want to round the numbers to just a few decimal points: The precision of 9 decimal places is uncalled for.

(N) Look at the Table 56.2.

Table 56.2

Multiple choice	a	b	c	d	e	f
#SEs (z)	1	1.96	2	2.5	2.58	3
p-value (two-tail)	0.317	0.050	0.046	0.012	0.010	0.003

(N1) Which of the entries are statistically significant using a 95% confidence level (a 0.05 alpha-level threshold)?

Answer: p-value ≤0.05 for b–f.

(N2) Which are statistically significant at the 99% confidence level?

Answer: p-value ≤0.01 for e–f.

(N3) Which are statistically significant at the 99.9% confidence level?

Answer: p-value ≤0.001 for none listed. Having a little over 3.0 for #SEs should do it.

(O) A surveyor surveys 100 random people in an evenly split community. Is it possible for the surveyor to get a random sample with 100% agree?

Answer: Yes, it is possible, but it is extremely, extremely unlikely. There are 2^{100} possible patterns of agree/disagree and only one of them is 100 agree (Chapter 3). The probability of 100 agree is $1/2^{100}$, which is about 10^{-30}, or about 0.000000000000000000000000000001.

(P) Suppose that a group of researchers conduct 1000 separate statistical tests. Alpha for all is 0.05. Beta for all is 0.20.

(P1) Let's say that all 1000 null hypotheses are actually true; how many type I and type II errors do we *expect*?

Answer: We expect 50 type I errors and 0 type II errors. Type II errors are only relevant when the null hypothesis is actually false.

(P2) Let's say that all 1000 null hypotheses are actually false; how many type I and type II errors do we *expect*?

Answer: We expect 0 type I errors and 200 type II errors. Type I errors are only relevant when the null hypothesis is actually true.

(P3) Let's say that half of the 1000 null hypotheses are actually false and half are actually true; how many type I and type II errors do we *expect*?

Answer: We expect 25 type I errors and 100 type II errors.

(P4) Suppose that the researchers instead set alpha to 0.01 and beta became 0.40, what is the answer for P3?

Answer: We expect 5 type I errors and 200 type II errors. By lowering alpha, we can expect fewer type I errors at the expense of more type II errors.

(P5) Suppose that all the researchers increase their sample sizes, how does the answer to P4 change?

Answer: If the researchers keep alpha at 0.01, then beta will decrease with the larger sample sizes. We still expect 5 type I errors but we expect fewer than 200 type II errors. How many fewer depends on how much the researchers increase their sample sizes.

(P6) After all is said and done, can we know how many type I and type II errors there actually are?

Answer: No.

57

For Part II

(A) Figure 57.1 is a simulation histogram of 1000 sample means. The samples come from a population with an unknown mean and variance. Does this look approximately like a normal distribution? What is your best rough estimate for the population mean?

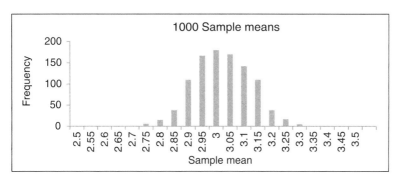

Figure 57.1

> *Answer:* Yes, it approximates a normal distribution. The unknown population mean is most likely close to 3.

(B) What interval holds about 95% of the results?

> *Answer:* Between about 2.8 and 3.2. About 25 of the sample means are less than or equal to 2.8, and about 25 are greater than 3.2. (Remember that with a histogram, each bar is the count of the sample means that are less

Illuminating Statistical Analysis Using Scenarios and Simulations, First Edition.
Jeffrey E Kottemann.
© 2017 John Wiley & Sons, Inc. Published 2017 by John Wiley & Sons, Inc.

than or equal to (≤) the number label for the bar itself and strictly greater than (>) the number label for the bar to the left.)

(C) So what does the standard error equal, approximately?

Answer: Since about two (1.96 rounded) standard errors on each side of 3 will hold 95% of the sample means (since it approximates a normal distribution), and the interval is about 3.0 ± 0.2, the standard error should equal about $0.2/2 = 0.1$.

(D) Figure 57.2 is a simulation histogram of 1000 sample mean differences. What is your best estimate for the difference between the actual population means?

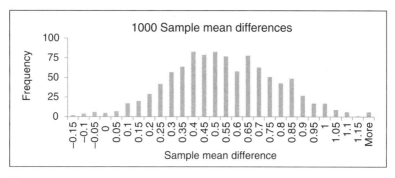

Figure 57.2

Answer: About 0.5

(E) What interval holds about 95% of the sample mean differences?

Answer: About 0.05 to 1.05.

(F) At the 95% confidence level (0.05 alpha-level) would you reject the null hypothesis that the two population means are equal?

Answer: Yes. Zero is outside the 95% confidence interval. There are very few sample means less than or equal to zero.

(G) Table 57.1 is from Chapter 26. Why do cases B and E have such a relatively small standard error and *p*-value?

Table 57.1

		s_F^2	s_M^2	n_F	n_M	Mean difference	Standard error (SE)	#SEs (z)	Two-tail p-value
A	Base case	3.1	2.9	100	100	−0.5	0.24495	−2.041	0.04123
B	Larger *n*	3.1	2.9	500	500	−0.5	0.10954	−4.564	0.00001
C	Smaller *n*	3.1	2.9	50	50	−0.5	0.34641	−1.443	0.14891
D	Larger s^2	8.0	8.0	100	100	−0.5	0.40000	−1.250	0.21130
E	Smaller s^2	1.0	1.0	100	100	−0.5	0.14142	−3.536	0.00041

Answer: Case B's relatively small standard error and *p*-value are due to its larger sample sizes. Case E's relatively small standard error and *p*-value are due to its lower sample variances.

(H) Let's say that the sample mean for females is 3.5 and for males is 3.4 with sample variances of 3.1 and 2.9. If we have a large enough sample size, can this small actual difference of 0.1 in the opinion averages be found to be a statistically significant difference?

Answer: Yes. To get to the 95% confidence level we would need a standard error of about 0.05, which fits twice into the actual difference of 0.1 (3.5 − 3.4). So, we need a value for sample size *n* so that the square root of $3.1/n + 2.9/n$ will give us 0.05. A sample size of 2400 does it. An *n* of 4000 will get us to about the 99% confidence level. (Feel free to do the arithmetic yourself to confirm this.)

General Answer: Statistical significance only indicates that our sample mean difference would most likely not have arisen by chance when the null hypothesis is true, not that the difference is significant in any practical sense – the latter is often called <u>practical significance</u>. With a large enough sample, the sample statistics will home in on any difference that might be in the population no matter how potentially trivial that difference might be.

There are methods that attempt to quantify practical significance via <u>effect size</u> statistics. In Part IV, *r*, *R* and R^2 are effect size statistics for relationships between variables. These effect size statistics can also be calculated for mean differences. (If you have not read Part IV, you might want to come back to this paragraph later.) For this example, and using an online effect size calculator, *r* equals about 0.03, which is very close to zero. This indicates that we may well have a case of statistical significance without practical significance. Nonetheless, we have to take into account the real-world context before making any pronouncements. Sometimes a small difference may be consequential.

Search online for "effect size statistics."

Here is another example of the distinction between statistical and practical significance using proportions. Say, there is a study showing that a certain dietary supplement lowers the risk of getting a certain ailment from 2 in a 1000 (0.002) down to 1 in a 1000 (0.001). If the sample size of the study is 30,000, then the p-value for the proportion difference between 0.002 and 0.001 is less than an alpha-level of 0.05. So, claims can be made that the supplement's *effect* is statistically significant and reduces the risk of getting the ailment by 50% $((0.002 - 0.001)/0.002)$. As evidence of *effect*iveness, however, many people would find the statistical significance and the relative risk difference of 50% to be misleading; people would also like to know the absolute risk difference of 2 in a 1000 versus 1 in a 1000, which is 0.2% versus 0.1% for an absolute difference of 0.1%.

Search online for "absolute versus relative risk."

(I) Table 57.2 has two p-values for each of the cases. One is based on using the z-distribution to determine the p-value and the other is based on using the t-distribution to determine the p-value. Why do the p-values for the t-distribution tend to be bigger than for the z-distribution?

Table 57.2

		s_F^2	s_M^2	n_F	n_M	Mean difference	SE	#SEs	z-distribution p-value	t-distribution p-value
A	Base case	3.1	2.9	100	100	−0.5	0.24495	−2.041	0.04123	0.04255
B	Larger n	3.1	2.9	500	500	−0.5	0.10954	−4.564	0.00001	0.00001
C	Smaller n	3.1	2.9	50	50	−0.5	0.34641	−1.443	0.14891	0.15210
D	Larger s^2	8.0	8.0	100	100	−0.5	0.40000	−1.250	0.21130	0.21277
E	Smaller s^2	1.0	1.0	100	100	−0.5	0.14142	−3.536	0.00041	0.00051

Answer: Because the t-distribution accounts for the additional uncertainty of using the sample variance (rather than the true population variance) in calculating standard error using the formula $\sqrt{(s^2/n)}$. The p-values are larger because of this additional uncertainty. This makes the most difference when we have small samples because we expect the sample variance to be less accurate with smaller samples.

(J) If we have sample sizes of $n = 1000$, will the p-value determined using the t-distribution be the same as the p-value determined using the z-distribution?

Answer: Yes, for all practical purposes.

(K) You have random samples of test scores for 100 females and 100 males. Your samples statistics are mean of 82 and variance of 20 for females, and mean of 80 and variance of 24 for males. Use a calculating tool and the below formula calculate the difference in terms of standard errors.

$$\frac{\text{sample mean}_F - \text{sample mean}_M}{\sqrt{\dfrac{s_F^2}{n_F} + \dfrac{s_M^2}{n_M}}}$$

Answer: The actual average score difference of 2 rescales to a standard error difference of a little over 3 standard errors.

(L) Would you say that the overall (population) average female score was higher than the overall average male score?

Answer: Yes, most likely, because 3 standard errors corresponds to a *p*-value less than 0.01. (With these fairly large sample sizes the *t*-distribution and the *z*-distribution are nearly the same.)

(M) But is an actual average score difference of 2 really "significant" in practical terms?

Answer: Again, just because something is *statistically* significant does not necessarily imply that it is *practically* significant. The only thing the statistical analysis is telling you is that there is most likely some degree of actual difference between the population averages for females and males. The actual difference in practice, however, might be trivial and without practical significance.

(N) Now let's explore some changes to the scenario (*italicized*). Say you have random samples of test scores for *400* females and *400* males. Your samples statistics are mean of *81* and variance of 20 for females, and mean of 80 and variance of 24 for males. Use a calculating tool and the above formula to calculate the difference in terms of standard errors. Estimate the *p*-value too.

Answer: Because of the larger sample sizes, the actual average score difference of only 1 rescales to a standard error difference of a little over 3 standard errors, which gives us a *p*-value of less than 0.01. This is a statistically significant difference at the same level we had in (L) above. An actual difference of 2 with samples sizes of 100 gives the same *p*-value as an actual difference of 1 with sample sizes of 400 (using sample variance of 20 and 24 for both scenarios).

58

For Part III

(A) In the X^2 (Chi-squared) chapters, we wondered if the community could be considered 25% Democrat, 25% Republican, and 50% Independents: This forms the basis for the null hypothesis. We have gathered a random sample of 120 community members, so we'll be testing for counts of 30, 30, and 60. Table 58.1 shows assorted scenarios (A–K) of what we might find in terms of the number of D, R, and I in our sample of 120. Take a moment to study the table. For each scenario, note its sample count values, the Chi-squared sum that results, and the p-value that is based on the magnitude of Chi-squared.

(A1) Which scenarios lead you to reject the null hypothesis with an alpha-level of 0.05?

Answer: Scenarios A and B and J and K. A surveyor is unlikely to get those sample count values for a community that is split 25, 25, and 50%.

(A2) Which scenarios cause you to reject the null hypothesis with an alpha-level of 0.01?

Answer: Now it is just scenarios A and K.

(A3) With an alpha-level of 0.05, which scenarios will lead you to *not* reject the null hypothesis that the community is split 25, 25, and 50%? Can you say that the community is split 25, 25, and 50% in those cases?

Answer: Scenarios C–I. No, you can not say that. You can only say that you have not ruled it out.

Illuminating Statistical Analysis Using Scenarios and Simulations, First Edition.
Jeffrey E Kottemann.
© 2017 John Wiley & Sons, Inc. Published 2017 by John Wiley & Sons, Inc.

Table 58.1 Chi-squared test for goodness of fit.

	D	R	I		
Testing for	30	30	60		
Sample count values				**Chi-squared**	***p*-Value**
Scenario A	40	40	40	13.33333	0.00127
Scenario B	38	38	44	8.53333	0.01403
Scenario C	36	36	48	4.80000	0.09072
Scenario D	34	34	52	2.13333	0.34415
Scenario E	32	32	56	0.53333	0.76593
Scenario F	30	30	60	0.00000	1.00000
Scenario G	28	28	64	0.53333	0.76593
Scenario H	26	26	68	2.13333	0.34415
Scenario I	24	24	72	4.80000	0.09072
Scenario J	22	22	76	8.53333	0.01403
Scenario K	20	20	80	13.33333	0.00127

(B) Table 58.2 shows the number of survey respondents that agree and disagree with a new national federal policy. Does the proportions that agree or disagree seem independent of what coast the respondents live on?

Table 58.2

	East Coast	West Coast	Count totals
Agree	800	300	1100
Disagree	200	500	700
Count totals	1000	800	grand count 1800

Answer: If you calculate the various proportions (or use a Chi-squared test for independence calculator), you will confirm what your eyes probably see: You can reject the null hypothesis of independence. East coasters tend to agree and West coasters tend to disagree. Agreement appears to depend on what coast the respondents live on.

(C) Table 58.3 shows the number of males and females enrolled in three different academic majors.

Table 58.3

	Major A	Major B	Major C
Male	25	30	45
Female	45	30	25

(C1) Use an online Chi-squared test for independence calculator to analyze the table. Do you reject the null hypothesis of independence between sex and major using an alpha-level of 0.05?

Answer: Yes, reject. The Chi-squared statistic is about 11.43 and p-value is about 0.0033.

(C2) In order to find the source of the dissimilarity between males and females and the majors they choose, do separate Chi-squared analyses for all the pairwise combinations. That is, test Male versus Female for Major A versus B, for Major A versus C, and for Major B versus C. Such follow up tests are called <u>post hoc</u> (Latin for "after this") tests.

Answer: A versus B gives Chi-squared of 2.70 and a p-value of 0.10; A versus C gives Chi-squared of 11.43 and a p-value of 0.0007; B versus C gives Chi-squared of 2.70 and a p-value of 0.10. So, it seems that Major A versus C is the primary source of the dissimilarity.

(C3) If you use an alpha-level of 0.05 for each of the three pairwise comparisons, does that give you a 0.05 alpha for *all three comparisons considered together?*

Answer: No. That is because if there is a 5% chance that a type I error could occur for *each* individual comparison, then there is more than 5% chance that a type I error could occur among *any* of the comparisons. By analogy, placing three individual bets that a tail will come up in *each* of three coin flips is not the same as placing one bet that a tail will come up in *any* of three coin flips.

Here is a simple way to determine how much lower to set the alpha-level for each comparison: Divide your desired overall alpha-level by the number of comparisons to be made. We have three pairwise comparisons, so if we want an overall alpha-level of 0.05, we should use $0.05/3 = 0.0167$ for each comparison. This is called the <u>Bonferroni Correction</u>. We'll see this issue again in (E2) below.

(D) You want to know if your community's population variances for incomes of males and females might be different. Samples are 100 random females and 100 random males. Table 58.4 shows nine different scenarios of what you might find with your samples.

Table 58.4 *F*-tests for differences in two sample variances.

Sample sizes 100 and 100				
	Sample variance male	Sample variance female	Variance ratio F	*p*-Value (one-tail)
Scenario A	4.0	4.0	1.0	0.5
Scenario B	4.5	4.0	1.12500	0.27950
Scenario C	5.0	4.0	1.25000	0.13432
Scenario D	5.5	4.0	1.37500	0.05740
Scenario E	6.0	4.0	1.50000	0.02247
Scenario F	6.5	4.0	1.62500	0.00825
Scenario G	7.0	4.0	1.75000	0.00290
Scenario H	7.5	4.0	1.87500	0.00099
Scenario I	8.0	4.0	2.00000	0.00033

(D1) In what value range does the variance ratio, F, first become statistically significant at the one-tailed alpha-level of 0.05?

Answer: Between scenarios D and E, with F between 1.375 and 1.5.

(D2) Using an alpha-level of 0.05, for which scenarios would you reject the null hypothesis of equal income variances for males and females in the overall community?

Answer: Scenarios E–I. These scenarios have *p*-values that are less than or equal to 0.05.

(E) Table 58.5 is a portion of the output for a one-way ANOVA produced by statistical analysis software. What's being investigated is the degree to which Democrats, Republicans, and Independents say that they keep up with current political news. 42 random members of each group answered a 1–7 scale survey questions, where higher numbers indicate a higher level of keeping current.

Table 58.5 ANOVA: single factor.

Groups	Count	Sum	Mean	Variance
D keep current	42	153	3.64	2.28
R keep current	42	189	4.50	2.50
I keep current	42	121	2.88	2.64
F	*p-value*			
11.13	<0.0001			

(E1) Is there a difference in the population means between any of the political affiliation groups?

Answer: Most likely. The *p*-value is less than 0.0001.

(E2) Can you tell which pairs of groups (D and R, D and I, R and I) have statistically significant differences in their sample means?
   Answer: ANOVA does not explicitly tell us. We can do individual post hoc *t*-tests to find out, but we will want to use a lower alpha-level. This is analogous to what we saw in C3 above. As we did in C3, to determine how much lower to set the alpha-level for each comparison we will simply divide the desired overall alpha-level by the number of comparisons to be made. We have three comparisons (D and R, D and I, R and I), so if we want an overall alpha-level of 0.05, we should use $0.05/3 = 0.0167$ for each comparison. Recall that this is called the Bonferroni Correction. There are many different correction methods to choose from. Do an online search for "post hoc multiple comparisons." Instead, or in addition, you can use the Tukey honestly significant difference (HSD) method to simultaneously assess all the pairwise comparisons.

(F) Table 58.6 shows a two-way ANOVA table based on random sample data, with the mean salaries for two companies broken out by sex. Does there seem to be an interaction effect?

Table 58.6

	Males	Females	Means
A Company	92,000	91,000	91,500
B Company	90,000	240,000	165,000
Means	91,000	165,500	grand mean 128,250

Answer: Yes, it seems that the average female salary is higher, but only when associated with Company B.

(G) Refer back to the scenarios given in Chapter 35 and determine which of the scenarios seem closest to each of the two ANOVA analyses shown in Table 58.7.

Table 58.7

Two-Way ANOVA #1				
Source of variation	*df*	**Variation**	*F*	*p*-**Value**
Male versus female means	1	256.887	237.876	<0.00001
Dem versus Rep means	1	115.509	106.961	<0.00001
Interaction	1	0.422	0.390	0.53330
Two-Way ANOVA #2				
Source of variation	*df*	**Variation**	*F*	*p*-**Value**
Male versus female means	1	0.093	0.087	0.768587
Dem versus Rep means	1	0.105	0.098	0.754558
Interaction	1	483.092	453.297	<0.00001

Answer: #1 is like scenario D, where both main effects are statistically significant and the interaction effect is not significant; #2 is like scenario E, where neither of the main effects are significant and the interaction effect is significant.

59

For Part IV

(A) Figures 59.1 and 59.2 are histograms showing simulation results of correlations between samples of *purely random numbers*. The first uses sample sizes of 30 pairs of numbers. The second uses larger sample sizes of 100.

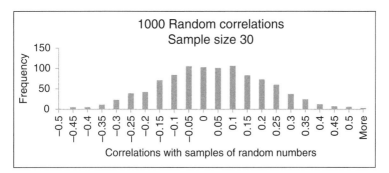

Figure 59.1

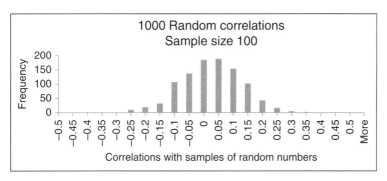

Figure 59.2

Illuminating Statistical Analysis Using Scenarios and Simulations, First Edition.
Jeffrey E Kottemann.
© 2017 John Wiley & Sons, Inc. Published 2017 by John Wiley & Sons, Inc.

(A1) Does it appear more likely to get a correlation of −0.2 (or less) purely by chance when you have a sample size of 30 or when you have a sample size of 100?

Answer: When you have sample size of 30. In general, the smaller the sample, the more likely it is to have larger correlations (negative or positive) by chance. This is similar to the fact that you are more likely to flip all heads when you flip a coin just a few times rather than many times.

(A2) Using real data with a sample size of 30, let us say you had a sample correlation of −0.25. Using the histogram in Figure 59.1 as a reference, would you expect your sample correlation to be statistically significant at the 0.05 alpha-level?

Answer: No. The area of the bars from −0.25 out to the left is fairly substantial. It looks like well over 50 random correlations are out in the left tail, giving a p-value greater than 0.05 (50/1000). That is one-tail. Doubling the one-tail area gives a two-tail p-value greater than 0.10 (100/1000).

(A3) Using real data with a sample size of 100, let us say you had a sample correlation of −0.25. Using the histogram in Figure 59.2 as a reference, would you expect your sample correlation to be statistically significant at the 0.05 alpha-level?

Answer: Yes. Looking at the histogram, it looks like about 10 of the 1000 sample correlations are less than or equal to −0.25; that is a one-tail p-value of 0.01. Two-tail is 0.02, which is less than 0.05.

(B) It has been conjectured that people who use social media more also watch television more. In other words, there are "media hungry" types of people and types that are less so. A survey of 186 young adults captured the self-reported time spent for social media and TV (both measured as average hours per day). The computed sample correlation between the two is 0.25 with a p-value of 0.00046.

(B1) Does the evidence support the conjecture?

Answer: Yes, statistically speaking. There is a statistically significant positive relationship suggesting that people who use social media more also watch television more. Notice that we are not claiming causality between the two, only association. It is statistically significant at the 0.001 alpha-level.

(B2) Does an r of 0.25, p-value <0.001, constitute a *practically* significant association?

Answer: Statistical significance tells you how likely it is that a given value for r could arise by chance when the null hypothesis is true. Practical significance, as we have seen, is a more common sense notion of what is meaningful in a particular context. As such, what is considered practically significant is dependent on context.

One common scheme categorizes correlation strengths as follows (using absolute values): 0.8 and above is "strong," 0.5–0.8 is "moderate," 0.2–0.5 is "weak," and below 0.2 is "negligible." However, such general categorizations are somewhat arbitrary and should not be applied too strictly. A correlation of 0.8 might be regarded as strong in a social science context where a complex human trait is being assessed with imprecise measurements, such as a survey using 7-point scales to measure peoples' attitudes. But, a 0.8 correlation might be regarded as weak in a physical science context where the traits of physical objects are assessed with precise measurements, such as sensitive instruments to detect particles and their movements in physics experiments.

(C) Table 59.1 shows correlations between college GPA, high school GPA, college entrance exam A score, and college entrance exam B score. The sample size is 130. Statistically significant (p-value<0.05) correlations are highlighted.

Table 59.1

	College GPA	High school GPA	Entrance exam A	Entrance exam B
College GPA	1.00			
High school GPA	0.37	1.00		
Entrance exam A	0.31	0.05	1.00	
Entrance exam B	0.21	−0.06	0.60	1.00

(C1) What correlates most strongly with college GPA?

Answer: High school GPA.

(C2) What does the relatively high positive correlation ($r = 0.6$) between entrance exams A and B indicate?

Answer: People who do well in one exam tend to do well in the other exam.

(C3) What impact would that have on a multiple regression model that uses high school GPA, entrance exam A, and entrance exam B together to predict college GPA?

Answer: The two entrance exams may be somewhat redundant predictors.

(C4) Will high school GPA be a redundant predictor?

Answer: No, because it is not significantly correlated with either of the entrance exam scores ($r = 0.05$ with A and $r = -0.06$ with B).

(D) Table 59.2 is a summary table of multiple regression results using the same data.

Table 59.2

	Coefficients	Standard error	t	p-Value
Intercept	0.92963	0.36458	2.54989	0.01197
High school GPA	0.38899	0.08373	4.64581	<0.00001
Entrance exam A	0.00110	0.00046	2.40758	0.01751
Entrance exam B	0.00041	0.00043	0.95384	0.34199
$R = 0.478$; Squared $= 0.229$; Adj. R-squared $= 0.210$; p-value<0.00001				

(D1) What does the Adjusted R-squared tell you?

Answer: 21% of the sample variance in college GPA is explained by the three predictors.

(D2) Is the overall regression model statistically significant?

Answer: Yes, in this context (unlike particle physics) a p-value less than 0.00001 would be considered statistically significant.

(D3) Which of the three predictors are statistically significant at the 0.05 alpha-level?

Answer: High school GPA (p-value<0.00001) and exam A (p-value $= 0.01751$).

(D4) Which of the three predictors are statistically significant at the 0.01 alpha-level?

Answer: Only high school GPA (p-value<0.00001).

(D5) Notice that entrance exam B is not statistically significant in this regression, even though we saw earlier that it has a statistically significant correlation with college GPA. Why is this?

Answer: As anticipated in Answers C2 and C3 earlier, the two entrance exams are somewhat redundant predictors. Collinearity is evident.

(D6) Can we say that entrance exam B is not predictive of college GPA?

Answer: No. In fact, the correlation results suggest that entrance exam B *is* predictive of college GPA. Entrance exam B simply does not contribute much to the predictive ability of our multiple regression model given the data we used.

(D7) What is the college GPA prediction formula?

Answer: CGPA$=0.92963+0.38899\times$HSGPA$+0.0011\times$EEA$+0.00041\times$EEB

(D8) What is the 95% confidence interval for the high school GPA coefficient?

Answer: The 95% confidence level is about two standard errors on each side. So, it is from $0.38899 - 2 \times 0.08373$ to $0.38899 + 2 \times 0.08373$.

(D9) Does the 95% confidence interval contain the value zero?

Answer: No, and there is no need to calculate the interval to answer this question. The high school GPA coefficient's p-value is less than 0.00001. The 95% confidence interval would contain zero only if the p-value was greater than 0.05.

(E) In the following example, let us look at a danger that often lurks in the tall grasses of correlation and regression analysis. Figure 59.3 is a scatterplot and Table 59.3 is the corresponding correlation table showing the relationship between how many hours a day students spend watching

TV and how many hours a day they study. The *p*-value is less than 0.000001. (This is hypothetical data that has been created for demonstration purposes.)

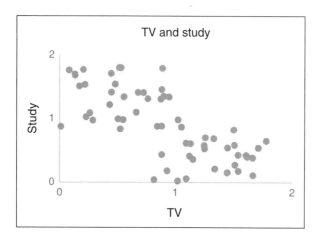

Figure 59.3

Table 59.3 Correlation of TV with study.

r	TV	Study
TV	1	
Study	−0.72	1

(E1) What does the value of *r* indicate?

Answer: More TV watching is associated with less studying, and the correlation is fairly strong.

(E2) Can you claim that TV watching *causes* people to study less?

Answer: No. There are many possible interpretations of the correlation. While TV watching can interfere with studying, studying can interfere with TV watching. Or, perhaps one or more overlooked variables are associated with studying and TV watching and are stealthily affecting the results. Further, perhaps an overlooked <u>confounding variable</u> is giving rise to a misleading, <u>spurious correlation</u> between studying and TV watching. One variable we might want to investigate is sex, with the thought that there may be systematic differences in TV watching and studying behaviors between males and females. Let us highlight females and males on the scatterplot, as shown in Figure 59.4, to see what we see.

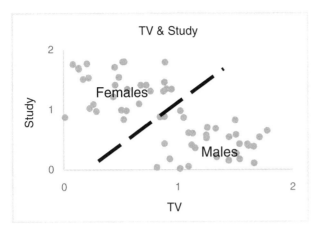

Figure 59.4

The overall negative slope of the data points that gives us a −0.72 sample correlation is due to the fact that females tend to watch TV less and study more, while males tend to watch TV more and study less. *But,* the data points within the female cluster and within the male cluster seem randomly scattered!

Tables 59.4a and b show the correlations between TV and study for females and males considered separately.

Those correlations are effectively zero. But, as we have seen, the overall correlation between TV and Study is −0.72. What is going on? Sex is a lurking variable that results in the spurious overall correlation of −0.71 between TV watching and studying. Tables 59.5a and b show the overall correlations of sex

Table 59.4 a and b Correlations of TV with study for females and males separately.

r	Female TV	Female study	r	Male TV	Male study
Female TV	1		Male TV	1	
Female study	−0.02	1	Male study	0.04	1

Table 59.5 a and b Overall correlations of sex with TV and sex with study.

r	Sex	TV	r	Sex	Study
Sex	1		Sex	1	
TV	−0.85	1	Study	0.83	1

with each of TV and study. *Both* correlations are strong. Sex is a confounding variable.

Let us analyze this using regression. The first results shown in Table 59.6 are for the simple linear regression model for study predicted by TV. Notice that TV is statistically significant. This is a spurious result that will lead us to make questionable conclusions.

Table 59.6 Simple regression of study with TV.

	Coefficients	Standard error	t	p-Value
Intercept	1.51562	0.08896	17.03666	<0.00001
TV	−0.67687	0.08651	−7.82394	<0.00001
$R = 0.717$; Square $= 0.513$; Adj. R-squared $= 0.505$; p-value <0.00001				

By failing to include sex in our analysis, we have unwittingly committed omitted variable bias. The next results, shown in Table 59.7, are for the multiple linear regression model for study predicted by TV and sex. Notice that TV is no longer statistically significant and sex is statistically significant.

Table 59.7 Multiple regression of study with TV and sex.

	Coefficients	Standard error	t	p-Value
Intercept	0.54931	0.18375	2.98940	0.00412
TV	−0.03140	0.13276	−0.23651	0.81389
Sex	0.78642	0.13773	5.70981	<0.00001
$R = 0.831$; R-square $= 0.690$; Adj. R-squared $= 0.680$; p-value <0.00001				

In the jargon, we would say that we are analyzing the relationship between study and TV while controlling for sex. (Equivalently, we could say while accounting for sex.) Given the multiple regression results, we find no significant relationship between study and TV (p-value $= 0.81$) when we control for sex. Controlling for potentially confounding variables in this way is commonplace in regression modeling.

Bottom Line: In general, correlations are not sufficient for asserting causality. A number of criteria have been proposed for doing so, such as demonstrating the existence of strong correlations as well as temporal ordering from causal variables to effect variables. And, as we have just seen, ruling out competing explanations is critical. There are advanced statistical techniques that can help in addressing such issues: Do an online search for "statistical methods for causal inference."

Note: There is a flipside to the above example. The scatterplot in Figure 59.5 shows a scenario where the overall correlation between TV and study is close to zero. However, when males and females are considered separately, as shown in Figure 59.6, the separate correlations between TV and study for males and for females are clearly negative. For this scatterplot, the overall correlation between TV and study is about −0.1, but the correlations between TV and study for males and for females calculated separately are each about −0.7.

Figure 59.5

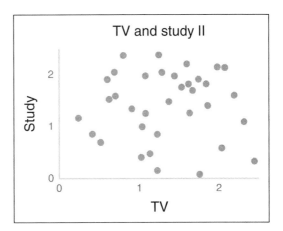

Figure 59.6

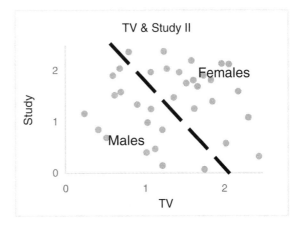

60

For Part V

(A) You conducted a small survey of 100 random adult members in your community with the following survey questions (assume everyone is perfectly willing to answer all your questions honestly). The nature of the numeric responses you received is also given for each survey item. The data appear to be unruly.

You want to do some statistical analysis. What would you do with the data?

Options: (a) omit outliers; (b) convert to nominal; (c) convert to ordered ranks (ordinal); (d) use a log transform.

(A1) How many people live in your household?
Most people answered with a value 1–6, but one wrote 99.

(A2) In a given year, how often do you go to the public library?
Over half of the people answered zero, the rest answered between 1 and 25.

(A3) How many days in a typical year are you out of town?
Answers are highly variable with a number of extreme values.

(A4) What is your annual income?
Values bunched up for low values and spread out for higher values.

Answers:
A1. May just want to omit the one outlier and make a note of it on your report.
A2. May want to recast this variable as a binomial: 0 for never go to the library, 1 for go to the library.
A3. May want to recast this as ordered ranks.

Illuminating Statistical Analysis Using Scenarios and Simulations, First Edition.
Jeffrey E Kottemann.
© 2017 John Wiley & Sons, Inc. Published 2017 by John Wiley & Sons, Inc.

A4. To test for differences, say between male and female, then may want to use the Mann–Whitney method that converts the data to ordered ranks. For regression, may want to use the log transformation of income.

(B) Table 60.1 lists ranks for females and males. Is there an "eyeball significant" ranking difference between males and females?

Table 60.1

F	F	F	F	F	M	M	M	M	M
1	2	3	4	5	6	7	8	9	10

Answer: Yes. The rank sum for females is as low as it can be and the rank sum for males is as high as it can be.

(C) Table 60.2 lists ranks for males and females. Is there an "eyeball significant" ranking difference between males and females?

Table 60.2

M	F	M	F	M	F	M	F	M	F
1	2	3	4	5	6	7	8	9	10

Answer: No, the rank sums will be almost equal.

(D) Table 60.3 lists ranks for females and males. Is there an "eyeball significant" ranking difference between males and females?

Table 60.3

F	M	F	M	F	M	F	M	F	M
1.5	1.5	3.5	3.5	5.5	5.5	7.5	7.5	9.5	9.5

Answer: No. Females and males are tied in pairs.

(E) Looking at the sampling simulation histogram in Figure 60.1, does the rank sum of 1900 appear to be outside the 95% confidence interval for the rank sums?

Answer: Yes, definitely.

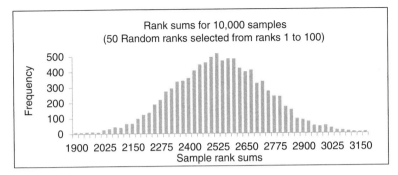

Figure 60.1

Appendices

A

Data Types and Some Basic Statistics

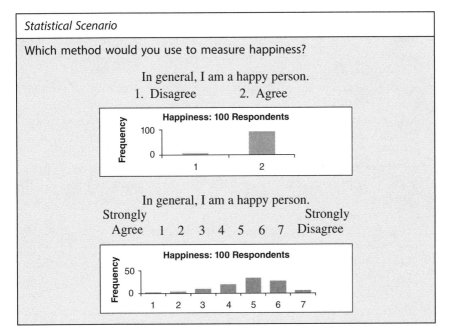

We often try to make sense of the world by trying to make sense of *data* about the world—naturally quantitative data such as heights, weights, incomes, and prices; quantitative indices of things such as inflation and unemployment; and, quantifications of qualitative things such as peoples' attitudes, beliefs, and opinions.

A fundamental aspect of data is its <u>measurement type</u>. The type of measurement determines how we can interpret the associated data, how we can manipulate and analyze it, and, in general, how much information it can relate. Figure A.1 outlines the commonly used types.

Illuminating Statistical Analysis Using Scenarios and Simulations, First Edition.
Jeffrey E Kottemann.
© 2017 John Wiley & Sons, Inc. Published 2017 by John Wiley & Sons, Inc.

Figure A.1

1) <u>Nominal</u> data, also called <u>categorical</u> data, have values that signify different categories. As examples, sex and political affiliation are both nominal variables: Sex is a <u>binomial</u> when there are only two categories allowed, "female" and "male"; political affiliation is a <u>multinomial</u> when there are more than two categories allowed, "Democrat", "Republican", and "Independent." The values of a nominal variable simply signify category membership.

   The basic computation we can do is <u>counting</u>: We can count the number of occurrences in the various categories and analyze the counts. Related to counts, we can also determine <u>proportions</u>. For example, based on counts of the number of people who say they "agree" with something versus the number who say they "disagree" we can determine the proportions of people who agree and disagree. We can then analyze these proportions. Statistical analysis methods for nominal variables are covered in Part I and Part III Chapters 28–32. Nominal variables also come into play in Part IV and make a brief appearance in Part V.

2) For <u>ordinal</u> data, we can assume that the data has a <u>rank ordering</u> but not that the consecutive ranks represent unit intervals: for example, if someone ranks their five favorite beverages, we can say that the second one is better liked than the fourth one, but we cannot say it is twice as liked nor that it is two units more liked.

   We can assign individual ranks, such as ranking individuals by grade-point average to determine their class rank, or ranking individuals by their annual income. We can also assign group ranks, such as assigning group rank of 1 for people with incomes of $0 to $19,999; 2 for incomes of $20,000 to $30,000; and so on. Statistical analysis methods using ordinal rank data are covered in Part V, although the book concentrates on nominal and scaled data.

3) <u>Scaled</u> data can be ratio scaled or interval scaled. For <u>ratio</u> data, ratios have meaning: for example, with price, saying something costs twice as much as something else makes sense. Same with heights and weights. For <u>interval</u> data, ratios are not fundamentally meaningful, but the distance between values is. For example, with degrees Celsius, we cannot meaningfully say that

40C is twice as warm as 20C and four times as warm as 10C, but we can meaningfully say that they are different by 20 and 30 degrees, respectively.

Because of their practical similarity, the ratio and interval measurement types are often lumped together and referred to as scaled measurements. In general, scaled measures reflect magnitudes, and so are more informative than ordinal measures. In many situations, scaled data enables the use of powerful methods for statistical analysis, which are covered in Part II, Part III Chapters 33–36, and Part IV. When scaled data is unruly, such as having several unusually extreme values, other methods covered in Part V can be used.

Some Basic Statistics (Primarily for Scaled and Binomial Variables)

(All of these statistics will be explained as they are introduced in the text.)

Statistics are computed values that serve to summarize a list of numbers in some way. Consider the following list of 11 numbers [1,2, 2, 3, 3, 3, 3, 3, 4, 4, 5].

The mode is the most frequently occurring value, which is 3. (This can also be used with nominal variables, where the mode is the category with the most members.)

The median is the middle value when the numbers are in sorted order. It is 3. (If there is an even number of values, then the average of the middle two values is used.)

The arithmetic mean is the numerical average, which is $33/11 = 3$. We will be referring to the mean extensively.

In math shorthand, the formula for the mean is

$$\frac{1}{n}\sum_{i=1}^{n}\text{value}_i$$

Σ is the Greek capital letter sigma, shorthand for sum; in this case, sum a list of n numbers. The sum is then divided by n via $\frac{1}{n}$.

The mode, median, and mean are all statistics of central tendency. The mean is most relevant for scaled numbers. Also, if we represent the two possible values of a binomial with the numbers 0 or 1, then the sum is the count of 1s and the mean is the proportion of 1s. For example, the mean and proportion of 1s for [0,1, 0, and 1] is 0.5 and the mean and proportion of 1s for [1,1, 0, and 1] is 0.75.

The variance is a statistic that reflects how disperse, or spread out, the number values are. Expressed in words, variance is the average squared difference between the values and the mean. (Squaring is done to make all the differences positive.) We will be referring to variance extensively.

In math shorthand, the fundamental formula for variance is

$$\frac{1}{n}\sum_{i=1}^{n}(\text{value}_i - \text{mean})^2$$

For binomials, when we represent the two possible values with the numbers 0 or 1, then this can be simplified to

$$p \times (1 - p)$$

where p is the proportion of 1s.

The square root of variance is called the <u>standard deviation</u>, which serves as another statistic reflecting how disperse, or spread out, the values are.

B

Simulating Statistical Scenarios

The type of <u>simulation</u> we will be doing is called Monte Carlo simulation—the use of random numbers to mimic realities involving chance. This appendix begins with step-by-step instructions and screenshots for constructing a basic statistical simulation in Excel. After that, it describes the nature of random variation evident in all simulations, then it overviews general guidelines for simulating statistical phenomena, and, finally, it gives specific instructions for reproducing the book's other statistical scenario simulations.

In cell A1 type = randbetween(0,1) then press Enter key on keyboard.

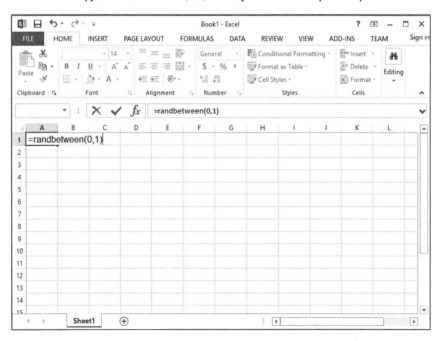

Illuminating Statistical Analysis Using Scenarios and Simulations, First Edition.
Jeffrey E Kottemann.
© 2017 John Wiley & Sons, Inc. Published 2017 by John Wiley & Sons, Inc.

You will now see either a 0 or a 1 in cell A1.

Press the function key F9 several times and you will see the value change like flipping a coin.

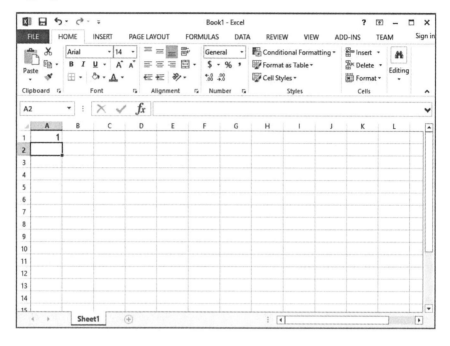

Copy A1 to B1 through J1 (B1:J1).
This will give you 10 coin flips in a row.

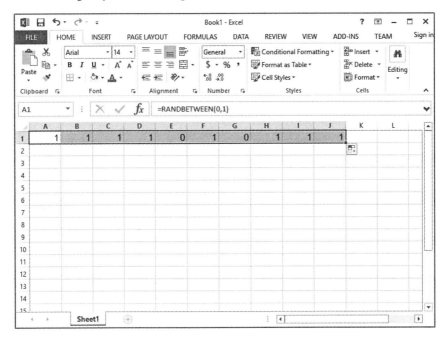

In cell K1 type = sum(A1:J1), then press Enter. This will give you the head count.

Press the function key F9 several times just to see what happens.

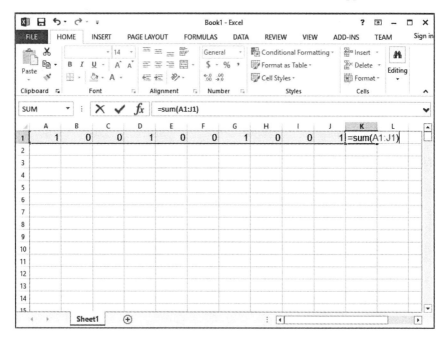

Copy cells A1:K1 down through row 1000.

This will give you 1000 independent repetitions of the 10-coin-flips scenario.

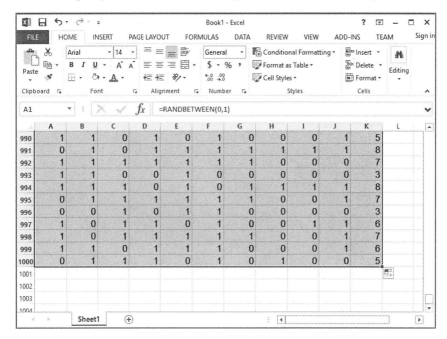

Now click Data tab, then Data Analysis on the far right, then click Histogram, then OK.

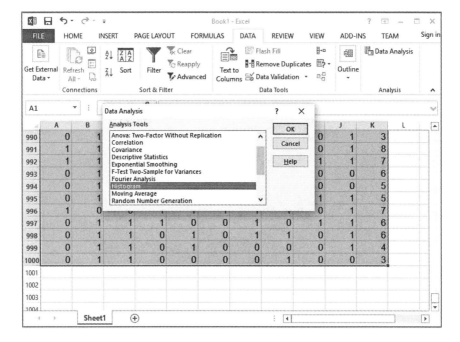

Type K1:K1000 in the dialog box's Input Range (or use the mouse to select those cells).

Check Chart Output.

Click OK.

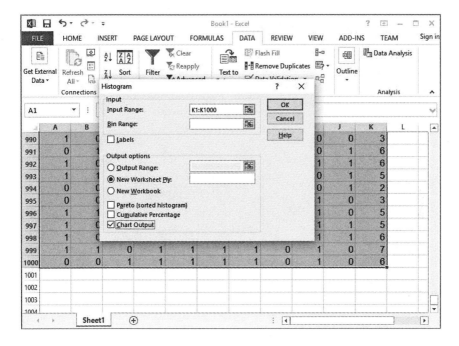

Now you will see the histogram on a new sheet.

You have now simulated 1000 repetitions of flipping a coin 10 times and counting the number of heads for each. The histogram summarizes the 1000 count outcomes. The histogram shows you the nature of the distribution of outcomes to expect for binomial (0 or 1) random processes—a bell-shaped distribution. Such phenomena are the topic Part I.

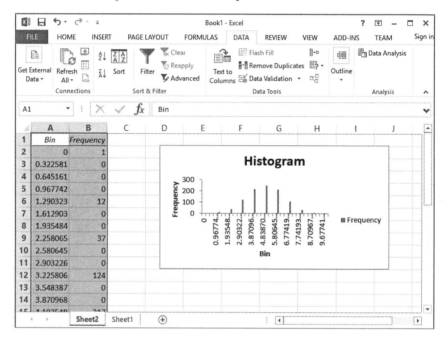

Notice that the histogram "Bin" label values chosen here automatically by the software are confusing. The actual values for the head counts are simply the integers 0–10, but the software has made bins like Frequency >0 and ≤0.322581, Frequency >0.322581 ≤645161, etc. (The automatic binning is usually better than this.)

The next screenshots and instructions show how to define your own bins for the horizontal axis of the histogram.

Return to your simulated-data sheet as shown below.

Type in the contents for the horizontal axis labels in the column M cells as shown. (Or, for a much easier way, look up "Autofill" in the spreadsheet help.)

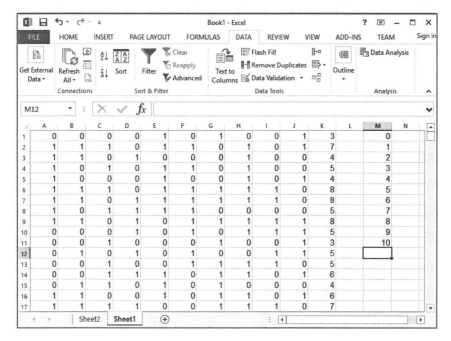

Now, as you did the first time around, click Data tab, then Data Analysis on the far right, then click Histogram, then OK.

Type K1:K1000 in the dialog box's Input Range (or use the mouse to select those cells).

Type M1:M11 in the dialog box's Bin Range (or use the mouse to select those cells).

Check Chart Output.

Click OK.

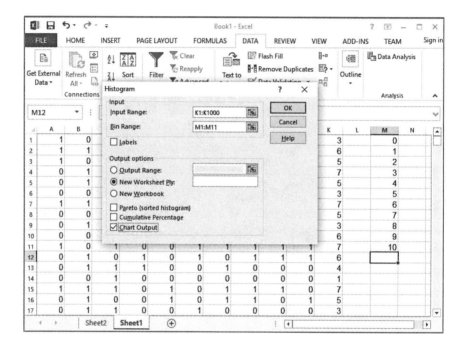

Now you will see the histogram with a nice, obedient horizontal axis.

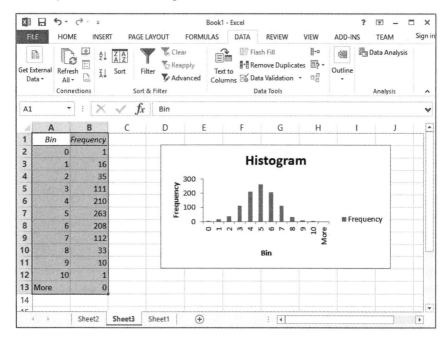

Random Variation

Throughout the book we will look at numerous histograms of simulation results that mimic reality, so it is important to understand a fundamental aspect of the simulations and the realities they represent: Each time we do a given simulation the results will be very similar, but slightly different. The two histograms shown in Figures B.1 and B.2 are the results of again and yet again simulating 1000 repetitions of flipping fair coins 10 times. You can see that the results are very similar, but slightly different. This would also happen if we really had 1000 people flipping fair coins.

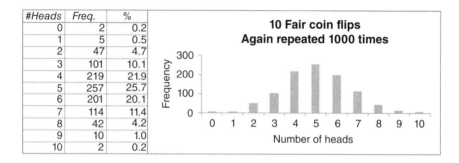

#Heads	Freq.	%
0	2	0.2
1	5	0.5
2	47	4.7
3	101	10.1
4	219	21.9
5	257	25.7
6	201	20.1
7	114	11.4
8	42	4.2
9	10	1.0
10	2	0.2

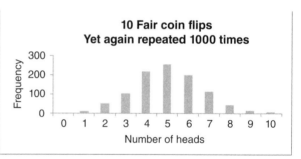

#Heads	Freq.	%
0	0	0.0
1	11	1.1
2	47	4.7
3	119	11.9
4	199	19.9
5	245	24.5
6	214	21.4
7	122	12.2
8	35	3.5
9	7	0.7
10	1	0.1

This is the nature of uncertainty—in simulation and in reality—when we randomly sample things, be it flipping coins or surveying randomly selected people for an opinion poll.

Note: Sometimes the simulation results histograms will come out unusually "lumpy" where some bars are quite a bit shorter or taller than expected. This happens just because of random chance. Since I want representative histograms for illustrations in the book, I redo the simulations in such cases. Or, sometimes I use 10,000 rather than 1000 rows. In general, the more repetitions (rows) you have, the more precise the overall results are, the closer to the theoretical expectation they will be, and the smoother the histograms come out.

General Guidelines

The number of columns that you use for your random numbers will be your sample size. For example, if you want to do the coin-flipping simulation for 30 coin flips at a time you will make 30 columns of RANDBETWEEN(0,1).

Optionally labeling columns: You can keep better track of the number of columns you are using if you use row 1 for labels. If you type DATUM1 in cell A1 and then copy it across, the spreadsheet software should automatically increment the labels to DATUM2, DATUM3, etc. in cells B1, C1, etc. (For more information on this automatic feature, look up "Autofill" in the spreadsheet help; it comes in especially handy for specifying histogram Bin Range values.) If you use row 1 for labels, you will need to check the "Labels" checkbox on the spreadsheet Histogram dialogue box.

The number of rows is the number of separate independent samples. If you want more than 1000 samples, just copy down as many rows as you like. I use 1000 most times and 10,000 when I make the histogram have a lot of separate Bins. In general, the more samples you have, the more precise the overall results are, the closer to the theoretical expectation they are, and the smoother the histograms come out.

If you use =AVERAGE instead of =SUM with the above example, you will get the proportion of heads instead of the number of heads for each sample. (You will also need to change the Bin label values in column M from 0, 1, 2, . . . 10 to 0, 0.1, 0.2, . . . 1)

Using =RAND() gives you a randomly selected *real* number between 0 and 1 uniformly distributed (equally likely). Notice that no arguments are needed, so we have the (). I rarely use this by itself, but I often use it as part of a formula with other functions, as shown below.

You can use =IF(RAND()>0.5,0,1) instead of RANDBETWEEN(0,1). This option will give you 50% tails and 50% heads on average by saying "if the 0-to-1 random real number is greater than 0.5, then assign the cell the value 0, otherwise assign it 1." *Note:* you could also use =IF(RAND()<0.5,0,1) instead.

Using =IF(RAND()>0.4,0,1) would simulate an unfair coin giving heads (1) 40% of the time. You can do the same to simulate surveying in a community in which 40% of the population agrees. This is used in Part I to simulate sampling in the community with population proportion of 0.4 agree (e.g., Chapter 10). *Note:* you could also use =IF(RAND()<0.6,0,1) instead.

Using =NORMINV(RAND(),0,1) will give you random numbers selected according to the probabilities of the standardized normal distribution introduced in Chapter 15 (the z-distribution, with mean of 0 and variance and standard deviation of 1). This is used in Chapter 28 to get the X^2 values. *Note:* you could also use =NORM.S.INV(RAND()) instead.

You can put in any values in place of the 0 and 1. For example =NORMINV (RAND(),7,2) will simulate random sampling from a population with population mean 7 and with population standard deviation 2 (variance 4). To specify the variance of 4 rather than the standard deviation of 2, you can use =NORMINV(RAND(),5,SQRT(4)). This approach is used in Part II, the second-half of Part III, and in Part IV.

Scenario-Specific Instructions

Each set of instructions tells you how to build row 1. When complete, copy the filled-in cells of row 1 down through row 1000, or as many rows as you wish. The last thing to do is make the histogram. If you use row 1 for labels (see above), then you will need to increment the row references given in the instructions by 1.

Chapter 5:
As given in the General Guidelines, the number of columns that you use for your random numbers will be your sample size. This chapter involves 2, 10, and 100 columns; each of the three constitutes a separate simulation.

Chapter 6:
As given in the General Guidelines, you can simulate different population proportions by making different simulations using different values for 0.5 in =IF (RAND()>0.5,0,1). In this chapter, we use 0.5, 0.25, and 0.1; each of the three constitute a separate simulation.

Chapter 7:
In cells A1 through AD1 put: =RANDBETWEEN(0,1) or you can use =IF (RAND()>0.5,0,1)
In cell AE1 put: =AVERAGE(A1:AD1)
Copy A1:AE1 down through row 1000.
Make a histogram of the cells AE1:AE1000.

Chapter 10:
In A1:CV1 have =IF(RAND()>0.4,0,1) *Note:* this gives on average 60% 0s, 40% 1s.
In CW1 have =AVERAGE(A1:CV1)
Copy A1:CW1 down through row 1000.
Make a histogram of the cells CW1:CW1000.

Chapter 12:
A1:GR1 have =RANDBETWEEN(0,1) or you can use =IF(RAND()>0.5,0,1)

In GS1 =AVERAGE(A1:CV1)
In GT1 =AVERAGE(CW1:GR1)
In GU1 =GS1-GT1
Copy A1:GU1 down as many rows as you wish.
Make a histogram of the difference cells in column GU.
You can make 10,000 rather than 1000 rows if you want to match the simulations in this chapter. However, using 10,000 rows may tax your computer and slow down its responsiveness. (This is because there will be $203 \times 10{,}000 = 2{,}030{,}000$ calculated cells.)

Chapter 14:
For #SEs, you can add a new column to the previous simulation as follows:
$GV1 = GU1/SQRT(0.25/100 + 0.25/100)$
Copy down
Make a histogram of the cells in column GV.

For simplicity, this uses the population proportions of 0.5 to calculate standard error. If you want to exactly mimic the chapter, make a column for the average of the two sample proportions and reference that cell for the $p \times (1 - p)$ calculation to replace the 0.25 shown above.

The frequency axis can also be changed to relative frequency by calculating the latter in an adjacent column and then copying & pasting-by-value those cells into the frequency column of the table produced by the Histogram command.

Chapter 15:
For *p*-values using the standard normal *z*-distribution with Excel, use the function =NORM.S.DIST(CellRef,1) where the CellRef cell holds the #SEs value. This function will give you the area under the normal curve to the left of #SEs. So, if #SEs is greater than zero, you need to use =1-NORM.S.DIST (CellRef,1) to get the area in the right tail.

You can handle both positive and negative #SEs in one formula by using =NORM.S.DIST(-ABS(CellRef),1)

The above will give a one-tail *p*-value. Multiplying the result by 2 gives the two-tail *p*-value.

If you want to draw the standard normal curve in a spreadsheet, put incremental #SEs values in column A starting in A1: $-4.5, -4.25, \ldots, 4.5$. Then type =1/ SQRT(2*PI()*EXP(-A1^2/2)) in cell B1 and copy it down. You will now have cells A1:B19 filled in. Finally, select those cells and do Insert Scatterplot with lines.

Chapter 16:
In A1:CV1 =IF(RAND()>0.5,0,1)
In CW1 =AVERAGE(A1:CV1)
Copy down and then make a histogram of the cells in column CW

Chapter 19:
In A1:AX1 =RANDBETWEEN(1,7)
In AY1 =AVERAGE(A1:AX1)
Copy down and then make a histogram of the cells in column AY.

Chapter 20:
Like Chapter 19 but have 200 columns (A1:GR1) of =RANDBETWEEN(1,7)

Chapter 21:
Like Chapter 19 but use =RANDBETWEEN(3,5)

Chapter 22:
In A1:CV1 =NORMINV(RAND(),4,1)
In CW1 =AVERAGE(A1:CV1)
Copy down and then make a histogram of the cells in column CW

Chapters 23 & 24:
Like Chapter 22 but use =NORMINV(RAND(),4.15,1)

Chapter 25:
In A1:CV1 =NORMINV(RAND(),3.0,SQRT(3.1))
In CW1:GR1 =NORMINV(RAND(),3.0,SQRT(2.9))
In GS1 =AVERAGE(A1:CV1)
In GT1 =AVERAGE(CW1:GR1)
In GU1 =GS1-GT1
Copy down and then make a histogram of the cells in column GU

Chapter 26: Use the same basic layout as Chapter 25. For the different sample sizes, use more or fewer columns of the =NORMINV cells. For different variances, replace the 3.1 and 2.9 in the =NORMINV formulas.

Chapter 27: Make four separate simulation sheets and put =NORMINV(RAND(),10,SQRT(4)) in 4, 7, 60, and 1000 columns. Have cells to the right, first with =AVERAGE second with =VAR.S and third with the spreadsheet formula for #SEs that is equivalent to $(\bar{x} - 10)/\sqrt{s^2/n}$. For each of the four sheets, and after copying down, make a histogram of the VAR.S cells and another histogram of the #SEs cells.

Chapter 28, making X^2 (Chi- squared) distributions: For individual cells in the simulation use =NORMINV(RAND(),0,1)^2 to get random z^2 values. When you have only one column of z^2, then just make a histogram of the z^2 column. When you have more than one column of z^2 values, make a =SUM column to the right and make a histogram of the SUM cells.

Chapter 33, making F-distributions: For the first simulation put =NORMINV (RAND(),10,SQRT(4)) in the 60 cells A1:BH1. In cell BI1 put =VAR.S(A1:AD1) and in cell BJ1 put =VAR.S(AE1:BH1). Last, in cell BK1 put =BI1/BJ1; this is the F cell. After copying down the row, make a histogram of the F cells in column BK. Make more or fewer NORMINV columns for the different sample sizes. Experiment with more simulations than are shown in the chapter. For example, try a scenario where the numerator has a sample size of 3 and the denominator has a sample size of 30.

Chapter 34: For the first simulation, use 30 columns of =NORMINV(RAND(),4, SQRT(1)) and =AVERAGE in the 31st column; for the second "hypothetical" situation use =NORMINV(RAND(),4,SQRT(30)); and for the third simulation use =NORMINV(RAND(),4,SQRT(9)). For each simulation, and after copying down, make a histogram of the AVERAGE cells.

Chapter 36: Random weights are =RANDBETWEEN(100,220)

Chapter 38: Put =NORMINV(RAND(),10,SQRT(4)) in the 240 cells A1:IF1. In cell IG1 put =CORREL(A1:DP1,DQ1:IF1). Copy down and then make a histogram of the CORREL cells. The second simulation is the same, but with fewer columns of random numbers.

The other simulations are a bit too involved to bother with.

Chapter 39: You can look at the distributions of correlation differences in the vicinity of zero with the following simulation. It is not shown in the book. You will have a total of 240 random number columns, all with the same =NOR-MINV function. Then have a column on the right that is the =CORREL of the first sixty columns with the second sixty columns. Have another column on the right that is the =CORREL of the third sixty columns with the fourth sixty columns. Add a final column on the right that calculates the difference between the two correlations. Copy the whole row down as many rows as you like and make a Histogram of the difference cells.

The other simulations are a bit too involved to bother with.

Chapter 41: For simulation of the slope coefficient b, use =NORMINV(RAND (),69,SQRT(16)) for the height values and =NORMINV(RAND(),164,SQRT

(504)) for the weight values. You will need 120 columns of each of these. On the right, make a column for =COVARIANCE.S(height cells, weight cells) and another column for =VAR.S(height cells) and last another column for b computed as the COVARIANCE.S cell divided by the VAR.S cell. Copy it all down for 1000 rows and do a Histogram of the b cells.

Chapter 50: For the simulation, in the 50 cells A1:AX1 have =RANDBETWEEN (1,100) and in cell AY1 have =SUM(A1:AX1). Copy down the row and make a histogram of the SUM cells. (Doing it this simple way allows for ties. Note that the formulaic method in Chapter 51 does not account for ties.)

Appendix C: Put =NORM.INV(RAND(),10,SQRT(4)) in cells A1:CV1. In cell CW1 put =AVERAGE(A1:CV1). Copy A1:CW1 down for 10,000 rows. In CX1 put =STDEV(CW1:CW10000). CX1 now holds the standard deviation of all 10,000 sample means. (You will *not* copy cell CX1 down; you only need the one standard deviation cell.) Press F9 multiple times. You will see that the value in CX1 will always be approximately equal to 0.2, which is the value for the standard error $\sqrt{(\sigma^2/n)} = \sqrt{4/100} = 0.2$

C

Standard Error as Standard Deviation

Variance and its square root, standard deviation, are general-purpose statistics for distributions, be it distributions of sample data or distributions of sample statistics. Below are the formulas for the *variance* and *standard deviation of sample data values* around the population mean, where n is the number of data values, x_i is the ith sample data value, and μ is the population mean.

$$\frac{1}{n}\sum_{i=1}^{n}(x_i - \mu)^2$$

$$\sqrt{\frac{1}{n}\sum_{i=1}^{n}(x_i - \mu)^2}$$

Next, let's look at the normal distribution that results from a sampling simulation of sample means for 10,000 random samples of size 100—see Figure C.1. The population mean μ is 10 and the population variance σ^2 is 4.

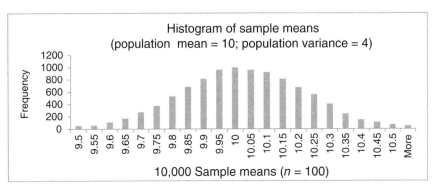

Figure C.1

Illuminating Statistical Analysis Using Scenarios and Simulations, First Edition.
Jeffrey E Kottemann.

Below are the analogous formulas for the *variance* and *standard deviation of sample mean values* around the population mean, where n is the number of sample means, $\bar{x}_i$ is the ith sample mean value, and μ is the population mean:

$$\frac{1}{n}\sum_{i=1}^{n}(\bar{x}_i - \mu)^2$$

$$\sqrt{\frac{1}{n}\sum_{i=1}^{n}(\bar{x}_i - \mu)^2}$$

Using this standard deviation formula with the 10,000 sample means from the simulation, we get 0.20.

Below is the formula for the *standard error of a sample mean*, where n is the number of sample *data* values, and σ^2 is the population variance:

$$\sqrt{(\sigma^2/n)}$$

Using this formula, we get $\sqrt{4/100}$ which also equals 0.20.

> Standard errors are standard deviations of
> sample statistics that have normal sampling distributions

In statistical inference, standard error is used when the sample statistic being considered can be assumed to be normally distributed, or can be transformed to be so. Examples include sample proportions (Part I), sample means (Part II), sample correlations and regression coefficients (Part IV), and sample rank sums (Part V).

D

Data Excerpt

This is the data behind the scenes for Part IV. This data is for illustration purposes only. The data is of real people, but numerous "extreme" values have been omitted, which was done to improve the clarity of scatterplots and analysis results.

Sex	Height	Weight
0	78	185
0	70	150
0	70	150
0	71	155
0	71	155
1	61	110
0	72	160
0	72	160
0	68	142
1	62	115
0	73	165
0	67	138
0	74	170
1	68	143
1	66	135
1	68	145
1	69	150
0	70	155
0	70	155
0	70	155
0	70	155

Illuminating Statistical Analysis Using Scenarios and Simulations, First Edition.
Jeffrey E Kottemann.
© 2017 John Wiley & Sons, Inc. Published 2017 by John Wiley & Sons, Inc.

274 | *Data Excerpt*

Sex	Height	Weight
1	60	110
0	71	160
0	71	160
1	71	160
1	61	115
0	72	165
0	72	165
0	73	170
1	65	135
0	68	149
0	77	190
0	67	145
1	68	150
1	68	150
0	70	160
0	70	160
1	64	133
0	71	165
1	61	120
0	72	170
0	73	175
1	64	135
0	65	140
1	65	140
0	67	150
0	68	155
1	62	128
0	71	169
0	70	165
0	70	165
0	70	165
0	71	170
0	71	170
0	71	170
0	71	170
0	72	175
1	62	130

Sex	Height	Weight
1	62	130
0	69	162
0	74	185
0	74	185
0	72	176
0	72	177
0	69	165
0	70	170
0	70	170
0	70	170
0	71	175
0	72	180
0	72	180
0	72	180
0	73	185
0	73	185
1	63	140
0	74	190
0	75	195
1	65	150
0	65	150
1	65	150
1	60	128
1	67	160
1	67	160
0	72	185
0	63	145
1	66	159
1	64	150
0	75	200
0	76	205
0	66	160
1	66	160
0	67	165
0	69	175
1	63	148
0	70	180

Sex	Height	Weight
0	70	180
1	65	158
0	72	190
1	62	145
0	73	195
0	63	150
0	74	200
0	65	160
0	68	175
0	70	185
0	70	185
0	69	181
0	72	195
0	75	210
0	69	185
0	75	213
0	71	195
0	72	200
0	65	170
0	76	220
0	62	160
0	65	175
0	70	200
0	70	200
0	72	210

E

Repeated Measures

Statistical Scenario—Weight Change

Let's say you want to test the effectiveness of a new diet. First, you get 30 random people to participate. You weigh them before they start the diet and then after 12 weeks on the diet.

The average weight change across all 30 participants is a loss of about 2 lbs.

Is this average weight loss statistically significant?

For this scenario, we cannot use the approach covered in Part II because the weights would just be lumped together into two groups: 30 before weights and 30 after weights. Instead, we need to assess the 30 *pairs* of before and after weights. For each pair, the weight before dieting is subtracted from the weight after dieting to give the difference in weight for each of the 30 individuals. Then, we can assess whether the *mean of the differences* is sufficiently far from zero. Table E.1 shows some of the dieting data along with the after-minus-before differences.

Table E.1 Example sample data.

Before	After	Difference
228	222	−6
228	230	2
223	219	−4
177	178	1
etc.	etc.	etc.

Illuminating Statistical Analysis Using Scenarios and Simulations, First Edition.
Jeffrey E Kottemann.
© 2017 John Wiley & Sons, Inc. Published 2017 by John Wiley & Sons, Inc.

The <u>Paired *t*-test</u> performs such an assessment. In a nutshell, after it calculates the individual differences and the mean and variance of the differences, it uses the variance of the differences and the number of differences to determine the <u>standard error of the differences</u>. Then, it determines how many standard errors separate the mean of the differences from zero, which in turn determines the *p*-value. Table E.2 shows the results of the Paired *t*-test provided by statistical analysis software.

Table E.2 Paired *t*-test results.

Paired *t*-test		
	Before	After
Mean	205.30	203.633
Variance	761.04	729.00
Observations	30	30
df (#differences minus 1)	29	
t (#SEs)	2.44983	
p-value, two-tail	0.02057	
t Needed for 95% confidence	2.04523	

At the 95% confidence level (alpha-level 0.05), we can claim that the new diet is successful. The average weight loss is less than 2 pounds; even so, it is statistically significant.

Figure E.1 shows corresponding simulation results for this scenario. Notice that very few of the average differences are zero or positive, corresponding to the small *p*-value (0.01 one-tail; 0.02 two-tail). Notice also the characteristic normal bell shape.

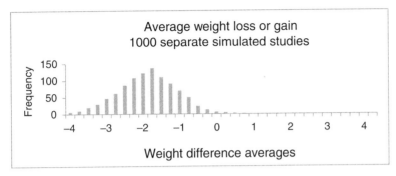

Figure E.1

Repeated measures ANOVA is an extension of the paired *t*-test. Mixed-effects models offer a more advanced approach. Do online searches using those terms for more information.

The ordinal ranking equivalent to the paired *t*-test is the Wilcoxon signed-ranks test (see Chapters 49–52 if necessary). Applied to the above example, the method would do the following: First, rank the differences ordered on their *absolute values*. Second, place the original signs onto the ranks. Third, if the sum of the signed ranks is relatively close to zero, then the evidence will not rule out the null hypothesis. As you probably guessed already, the distribution of signed-rank sums is normal, and there is a formula for the standard error of signed-rank sums.

F

Bayesian Statistics

Bayesian statistics provides a special method for calculating probability estimates and learning about unknown population statistics. To explore its workings, let's use the framework introduced in Chapter 1. This time we will use a statistical scenario involving medical diagnosis rather than courts of law. Table F.1 shows the various possible conditions and outcomes of a diagnostic test for a fictional disease, Krobze.

Table F.1 Scenario structure.

test result / Unknown truth	Test negative	Test positive
Don't have Krobze	true negative	false positive (type I error)
Do have Krobze	false negative (type II error)	true positive

Bayesian inference can be used to estimate conditional (if . . . then . . .) probabilities like:

1) If you test negative, then what is the probability that you really don't have Krobze?
2) If you test positive, then what is the probability that you really do have Krobze?

The first involves the test negative column: we need to calculate the probability of a true negative divided by the sum of the probabilities of true and false negatives. The second involves the test positive column: we need to calculate the probability of a true positive divided by the sum of the probabilities of true and false positives.

In order to calculate these, we need to know the probability of having the disease in general—that is, how common Krobze is in the population. We also

Illuminating Statistical Analysis Using Scenarios and Simulations, First Edition.
Jeffrey E Kottemann.
© 2017 John Wiley & Sons, Inc. Published 2017 by John Wiley & Sons, Inc.

need to know how reliable the diagnostic test is. Listed below is this required information, which is then used to fill in Table F.2.

1 out of 1000 have Krobze, for a probability of *0.001* of having Krobze and *0.999* of not.

For the diagnostic test reliability we are given the following information:

For people who don't have Krobze, 90% (*0.9*) test negative (true negative) and 10% (*0.1*) test positive (false positive).

For people who do have Krobze, 80% (*0.8*) test positive (true positive) and 20% (*0.2*) test negative (false negative).

Table F.2 Probabilities after a single testing.

Test result / Krobze?	Test negative	Test positive
Don't have Krobze (0.999 of the population)	true negative 0.999 × 0.9 = 0.8991	false positive 0.999 × 0.1 = 0.0999
Do have Krobze (0.001 of the population)	False negative 0.001 × 0.2 = 0.0002	True positive 0.001 × 0.8 = 0.0008
Column probability sums	0.8993 (89.93% will test negative)	0.1007 (10.07% will test positive)

After studying Table F.2, we can answer the questions posed above:

1) If you *test negative*, what is the probability that you *don't* have Krobze? For this we use the test negative column.
   True negative/true and false negatives = 0.8991/0.8993 = 0.999778, which is nearly 100%.

2) If you *test positive*, what is the probability that you *do* have Krobze? For this we use the test positive column.
   True positive/true and false positive = 0.0008/0.1007 = 0.007944, which is less than 1%. (It is this low because Krobze is so rare that the false positives overwhelm the true positives.)

Bayesian terminology:

The two don't/do have Krobze cells show <u>prior probabilities</u>.
The four true/false negative/positive cells show <u>joint probabilities</u>.
The two column probability sums cells show <u>marginal probabilities</u>.
The two calculated solutions 0.999778 and 0.007944 are <u>posterior probabilities</u>.

Many people feel that all of this makes better sense when we use frequencies rather than probabilities, let's us do it that way too.

Consider 10,000 random people. Below are the frequencies we expect, which are then used to fill in Table F.3.

9990 will not have Krobze (999 out of 1000 don't have it) and of those,
  8991 (90% of 9990) test negative (true negative)
  999 (10% of 9990) test positive (false positive)
10 will have Krobze (1 out of 1000 have it), and of those
  8 (80% of 10) test positive (true positive)
  2 (20% of 10) test negative (false negative)

Table F.3 Using frequencies instead of probabilities.

Test result Krobze?	Test negative	Test positive
Don't have Krobze (9990 out of 10,000)	True negative $9990 \times 0.9 = 8991$	False positive $9990 \times 0.1 = 999$
Do have Krobze (10 out of 10,000)	False negative $10 \times 0.2 = 2$	True positive $10 \times 0.8 = 8$
Column frequency sums	8993 (8993 will test negative)	1007 (1007 will test positive)

1) If you *test negative*, how likely is it that you *don't* have Krobze?
   $8991/8993 = 0.999778$; same as above.
2) If you *test positive*, how likely is it that you *do* have Krobze?
   $8/1007 = 0.007944$; same as above.

The Bayesian method is especially useful because it can be used in succession to update probability estimates. Let's say you tested positive the first time and want to have another type of test (with the same diagnostic reliability) performed. Table F.4 shows the conditions and outcomes for the second test. The posterior probability of 0.007944 based on your first positive test now becomes the prior probability for the second test.

Table F.4 Second test following the first positive test result.

second test result Krobze? (given first positive test)	Test negative	Test positive
Don't have Krobze $1 - 0.007944 = 0.992056$	true negative $0.992056 \times 0.9 = 0.89285$	false positive $0.992056 \times 0.1 = 0.099206$
Do have Krobze 0.007944	false negative $0.007944 \times 0.2 = 0.001589$	true positive $0.007944 \times 0.8 = 0.006355$
Column probability sums	0.894439	0.105561

Say that you test negative the second time. The probability that you *don't* have Krobze given that the second test is negative:

0.89285/0.894439 = 0.998223; nearly 100%

Instead, say that you again test positive. The probability that you *do* have Krobze following the second positive test result is

0.006355/0.105561 = 0.060202; about 6%

Keep in mind that Krobze is very rare. So, false positive test results tend to overwhelm the test positive column: that is largely why we got 1% after the first positive test and 6% after the second positive test. Test reliability matters too, as we will see next.

Let's say that instead, the doctor orders a much more reliable (and much more expensive) test the second time. It detects both true negatives and true positives 99% of the time. Table F.5 covers this situation.

Table F.5 Second super-test following the first positive test result.

second super-test result Krobze? (given first positive test)	Test negative	Test positive
Don't have Krobze 1 − 0.007944 = 0.992056	true negative 0.992056 × 0.99 = 0.982135	false positive 0.992056 × 0.01 = 0.009921
Do have Krobze 0.007944	false negative 0.007944 × 0.01 = 0.000079	true positive 0.007944 × 0.99 = 0.007865
Column probability sums	0.982214	0.017786

Say that you test negative the second time with the super-test. The probability that you *don't* have Krobze given the first positive test result and the second negative super-test result is

0.982135/0.982214 = 0.999920; nearly 100%

Finally, say that you test positive the second time with the super-test. The probability that you *do* have Krobze given the first positive test and the second positive super-test is

0.007865/0.017786 = 0.442202; about 44%

A Note on Priors

With the above analyses we were extremely fortunate to know that, in general, 1 in 1000 people have Krobze. That gave us our initial prior probabilities (0.999 and 0.001) that we needed to get started. What if we don't know these initial prior probabilities? One alternative is to use expert opinion, which is somewhat subjective. Another is to use what are called "noninformative priors." For example, if we have no idea what the initial prior probability of Krobze is, we could use the noninformative priors of equal probabilities for don't have Krobze (0.5) and do have Krobze (0.5). As you can imagine, we will get quite different results using priors of 0.5 and 0.5 instead of 0.999 and 0.001. Determining valid priors is important, and it can be tricky.

Getting More Sophisticated

More sophisticated Bayesian analysis involves entire *distributions* rather than specific values. For example, let's say we are trying to estimate a population mean. Figure F.1 shows (1) the prior distribution for the mean estimate proposed by an expert, (2) the mean estimate distribution based on sample data, and (3) the resulting posterior distribution for the mean estimate derived via Bayesian updating methods. As usual, uncertainty due to sample size and variance will affect the spread of the distributions.

Figure F.1

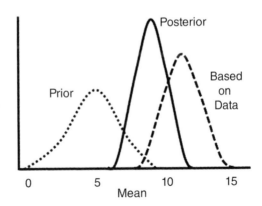

From the posterior, we can see that the most likely estimate for the population mean is about 9. We can also determine the interval containing 95% of the posterior's area: 7–11. This is a Bayesian <u>credible interval</u>, which is subtly different than a classical confidence interval. *Credible interval:* Given our sample data, there is a 95% chance that the true population mean falls within the 95% credible interval; valid priors and distributional assumptions need to be incorporated into the analysis. *Confidence interval:* 95% of the 95% confidence intervals derived from sample data will contain the true population mean; no priors are needed and the statistical methods themselves embody distributional assumptions. Do an Internet search with "credible versus confidence interval" for more on this. And do an Internet search with "Bayesian versus frequentist statistics" for more on the general distinctions.

Note: If you look back to the last section of Chapter 17 on type I and type II errors, you'll see that we used basic Bayesian analysis to calculate the false discovery rate.

G

Data Mining

Data mining aims to discover and make use of patterns and relationships that are manifest in data. There are many different methods that can be used for data mining, and linear regression is one. In a data mining context, we could use linear regression to build a model to predict numbers of interest. Then, once the model is built using a set of cases, we would assess how well the model can predict a new set of cases. Further, we'll also want to compare the model's performance to that of models constructed using other predictive analytics methods, such as neural networks and decision trees. Let's go through this process a step at a time.

We will no longer dwell on p-values, but instead we'll focus on measures of fit such as Adjusted R-squared. Further, we will not dwell so much on the adjusted R-squared we get when we build a model, but instead we'll focus more on the adjusted R-squared we get when we use the model to predict new cases.

Below is the regression equation we developed in Chapter 43. I will call it "The Model."

$$\text{Weight} = -100.24 + 3.87\text{Height} - 9.20\text{Sex}; \ R = 0.82; \ R^2 = 0.68; \ \text{Adj } R^2 = 0.67$$

The above adjusted R-squared value of 0.67 tells us how well the model predicts weights of the people whose data were used to build the model in the first place. But, in data mining applications, we want to test how well the model will predict the weights of new people, that is, people whose data were not used to build the model.

Luckily, I happen to have some new data—a dataset of height, sex, and weight for 193 new people. How can we use this new data to validate The Model? We can use The Model's regression equation to predict the weights of the new people and then see how well the predictions match the actual weights of the new people. To illustrate, shown in Table G.1 is a section of spreadsheet with six new cases. The left three columns hold the new data. The right column has The Model calculate predicted weight using the new height and gender data.

Illuminating Statistical Analysis Using Scenarios and Simulations, First Edition.
Jeffrey E Kottemann.
© 2017 John Wiley & Sons, Inc. Published 2017 by John Wiley & Sons, Inc.

Table G.1 New data and predictions.

Height	Gender	Weight	The Model's predicted weight
70	0	180	171
73	0	205	182
62	1	126	131
73	0	183	182
66	1	153	146
60	1	120	123

Header note: Data on 193 new people spans Height, Gender, Weight.

Using the first case as an example, we have the actual weight of 180 and The Model's predicted weight of

$$171 = -100.24 + 3.87 \times 70 - 9.20 \times 0$$

We can calculate a new R-squared for The Model by taking the squared correlation between the column of 193 actual weights and the column of 193 predicted weights. Doing so yields an R-squared of 0.45 and an adjusted R-squared of 0.44. This adjusted R-squared gives us a better idea of how well our model will work when it is used to predict new cases, and we can compare different models this way.

To adopt this type of model validation strategy, you don't have to wait for new data to show up. You can randomly split your original dataset into two subsets and use one subset to build the model and the other subset to validate the model. In the jargon, one subset is used for model training (building) and the other is used for model validation. A similar process of model training/ validation is commonly used in data mining to evaluate prediction models, whether the models are built using linear regression or using methods such as neural networks and decision trees.

Next, let's take a visual tour of several data mining methods using the two scatterplots of *training data* shown in Figure G.1a and b. The first scatterplot depicts a linear relationship and the second scatterplot depicts a nonlinear relationship. We want a model that can best predict Y given X for the first scenario, and another model that can best predict Y given X for the second scenario. Hypothetically, the first scatterplot might portray working-age adults with X being age and Y being deposits to personal savings, and the second scatterplot might portray all adults with X again being age and Y again being deposits to personal savings (in retirement, savings deposits fall).

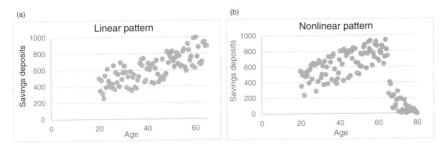

Figure G.1

Using linear regression with the training data, we would get the regression lines shown in Figure G.2a and b. The first line is a good fit. The second line is a bad fit. When evaluating new cases based on these linear models, we would likely find that the first model works well, but that the second model does not. Validation data will help us assess that.

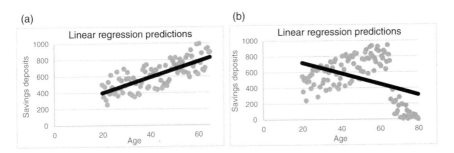

Figure G.2

Instead, we could fit a more flexible prediction line to the training data with a nonlinear model constructed using <u>polynomial regression</u>. Shown in Figure G.3a and b are what we get.

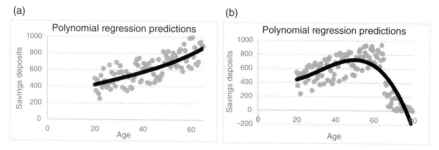

Figure G.3

For the first scenario, the polynomial prediction line is nearly linear. For the second scenario, the polynomial prediction line is

$$\text{Savings deposits} = 961.763 - 59.949\text{Age} + 2.004\text{Age}^2 - 0.018\text{Age}^3$$

We can enjoy even more flexibility with a *network* of nonlinear functions constructed using a <u>neural network</u> method. Shown in Figure G.4a and b is what we get.

Looking at the diagrams, the predictive accuracy of the neural network seems comparable to that of the regression models for the first scenario. On the other hand, the neural network does better for the second scenario.

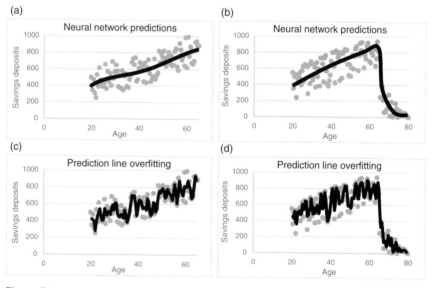

Figure G.4

Neural networks are constructed by repeatedly exposing them to the training data, and, because they are so flexible, there is a danger that the network may be "over trained" and become unduly specialized to idiosyncrasies in the training data. Such a situation is called overfitting. Figure G.4c and d illustrate extreme overfitting. The validation process should alert us to overfitting because the fit of the model to validation data would degrade.

For our next approach, let's successively subdivide the training data into subregions, as shown in Figure G.5a and b (the thickness of the lines indicates the level of subdivision, major to minor). Each subregion will become associated with a Y^* value that is the average of the Y values in that subregion (stars in the diagram). For new cases, the value of X will correspond to one of the subregions and the prediction will be set equal to that subregion's Y^* value.

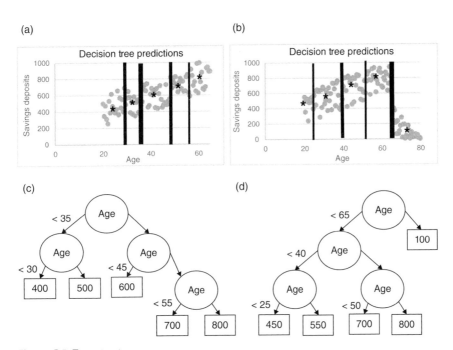

Figure G.5 Tree structure.

Decision tree methods successively subdivide the training data in this way. Tree structure renditions of Figures G.5a and b are shown in Figure G.5c and d. Beginning at the top decision node in a tree, the value for age determines whether to follow the left or right branch at each successive level in the tree. For example, let's say that we have a new case with age equal to 33. Using tree

Figure G.5c, we first follow the left branch because 33 is less than 35. At the next level, we follow the right branch because 33 is not less than 30. This branch then ends with a prediction of 500 for savings deposits.

We can potentially improve decision tree models by making further subdivisions in order to make each subregion more homogenous. But, if we make too many small subregions based on the training data, the tree not only becomes quite large it also becomes overly specialized to the training data—overfitting. (An illustration of decision tree overfitting is shown later.) This is analogous to what can occur with neural networks, and again, the validation process should alert us to overfitting.

Given all the modeling results illustrated above for the training data, all the data mining methods show potential for the first scenario. For the second scenario, the neural network and decision tree methods show the most potential. In practice, extensive model validation within and across candidate models would need to be performed before finalizing the model(s) and choosing among them. Single models may be chosen, or ensembles of models may prove more effective. (Ensemble modeling is the process of using two or more models and combining their predictions—by averaging, for example.) After that, ongoing model reevaluation and recalibration would need to be conducted while they are in use.

In addition to methods for developing prediction models, related methods can be used to develop models for classification, where the Y variable is binomial or multinomial. For example, Figure G.6a shows the results of applying decision tree logic to subdivide a data-space involving two scaled X variables and a binomial Y variable. Figure G.6b shows its logical decision tree structure.

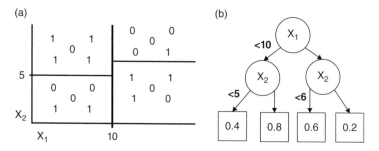

Figure G.6 Dataspace partitioning and the corresponding decision tree structure.

The Y^* value for each subregion is calculated, as before, to be the average of the values in the subregion. The Y^* values will be between 0 and 1 and can be used as probability estimates. Using a probability cutoff of 0.5, new cases corresponding to a subregion with Y^* less than 0.5 will be classified as 0 and otherwise will be classified as 1.

As noted earlier, an objective when building decision trees is to have the subregions relatively homogenous, but without making the subregions so small that we get unwanted overfitting to the training data. Figure G.7 shows overfitting. It seems unlikely that these overly exacting subregions in the training data will correspond well to new cases. Model validation helps assess that.

Figure G.7 Overfitting to the training data.

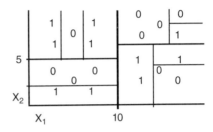

Prediction and classification methods focus on determining Y values given sets of X-values. Such modeling is called <u>supervised learning</u> in the sense that the Y variable supervises the role that the various X variables will have in the model. Other data mining methods concern situations where there is no Y variable involved. Since there is no supervising Y variable, this is called <u>unsupervised learning</u>. Two popular methods we'll glimpse at next are <u>cluster analysis</u> and <u>affinity analysis</u>.

Cluster analysis groups cases into clusters (subgroups) that have similar values for a set of X variables. Cluster analysis views relationships in terms of the distances between cases and between clusters. In a nutshell, it determines clusters that have small distances between cases within each cluster, but large distances between the clusters. Figure G.8a and b depict the results of cluster analysis involving two X variables. In Figure G.8a, the data-analyst specified that two clusters be formed. In Figure G.8b, the data-analyst specified that three

Figure G.8

clusters be formed. This data might represent customers in terms of their household incomes and how often they go shopping, and the clustering might be used to characterize different categories of shoppers.

Affinity analysis views relationships in terms of co-occurrence. For example, given a large number of individual shopping receipts stored electronically in a grocery store's database, affinity analysis can derive facts such as: 80% of shopping trips where cheese was purchased, crackers were also purchased; 30% of shopping trips where cheese *and* grapes were purchased, crackers were also purchased. You have undoubtedly been the recipient of such analyses when, as an online shopper, you received notice that people who purchased what you purchased also purchased certain other things too.

There are quite a few software packages available that support data mining in one way or another. Their capabilities can include an array of modeling methods to choose from as well as facilities to support the training/validation process. Long-established statistical analysis packages—such as SAS and SPSS—have integrated data mining capabilities available. Do an online search for "data mining tools and techniques."

Index

Illuminating Statistical Analysis Using Scenarios and Simulations, First Edition.
Jeffrey E Kottemann.
© 2017 John Wiley & Sons, Inc. Published 2017 by John Wiley & Sons, Inc.